PRAISE FOR

MW00748539

At once alarming and inspiring, *The Green Chain* reminds us why we should all love trees. Leiren-Young's profiles of activists, loggers and politicians give readers true insight into the complex world of forestry.
　　　　　—Vanessa Farquharson, author of *Sleeping Naked is Green*

Mark Leiren-Young is a Canadian national treasure: writer, journalist, filmmaker, songwriter, stand-up comedian and now podcaster. His new book, *The Green Chain*, reflects his extraordinary commitment to the environment in general and trees in particular—because if they go, we go. I hope this book brings him the global audience he deserves.
　　　　　—Jon Cooksey, writer-director-star of *How to Boil a Frog*

Mark Leiren-Young is that rare writer who actually can see both the forest and the trees at the same time. In this fascinating book Mark talks to people in both the industry and the environmental movement with a curiosity that brings out the true essence of forestry.
　　　　　—Bill Tieleman, columnist, *24 Hours* newspaper

Green Chain is a fascinating, compassionate and humorous account of the characters that make forest issues in BC so lively and relevant for all of us. One can't help but be enriched by the people in this book and the stories they tell.
　　　　　—Bruce Lourie, co-author of *Slow Death by Rubber Duck*

I love trees. And I love a good conversation. All change starts with conversations, and right now the world needs a lot more of them. *The Green Chain* book will spark a lot of important conversations about the future of forests and forestry in Canada and around the world.
　　　　　—Leif Utne, journalist, activist, online community builder

the GREEN CHAIN

the
GREEN CHAIN
nothing is ever clear cut

MARK LEIREN-YOUNG

VICTORIA • VANCOUVER • CALGARY

Heritage House Publishing Company Ltd.
#108–17665 66A Avenue
Surrey, BC V3S 2A7
www.heritagehouse.ca

Library and Archives Canada Cataloguing in Publication

Leiren-Young, Mark
  The green chain: nothing is ever clear cut / Mark Leiren-Young.

Includes screenplay of The green chain.
ISBN 978-1-894974-89-9

  1. Forests and forestry—Canada. 2. Forests and forestry—Economic aspects—Canada. 3. Forests and forestry—Environmental aspects—Canada. I. Title.

SD145.L45 2009   333.750971   C2009-904507-9

Edited by Heather Sangster
Proofread by Lenore Hietkamp
Cover design by Jacqui Thomas
Book design and layout by Pete Kohut
Front-cover image by Harry Bardal Design
Textured-paper image on back cover by Peter Zelei/iStockphoto

Mixed Sources
Cert no. SW-COC-001271
© 1996 FSC
FSC

The interior of this book was produced on 100% post-consumer recycled paper, processed chlorine free and printed with vegetable-based dyes.

BRITISH COLUMBIA
ARTS COUNCIL
Supported by the Province of British Columbia

Canada Council   Conseil des Arts
for the Arts   du Canada

Heritage House acknowledges the financial support for its publishing program from the Government of Canada through the Book Publishing Industry Development Program (BPIDP), Canada Council for the Arts, and the province of British Columbia through the British Columbia Arts Council and the Book Publishing Tax Credit.

Printed in Canada

"Everyone has their own personal tree to climb."
—Julia Butterfly Hill

For the producers, cast and crew of *The Green Chain*. It was a joy and an honour to climb the tree with you.

CONTENTS

INTRODUCTION

I love trees.

When I wrote the script for my movie *The Green Chain*—a collection of seven fictional interviews about the people behind the issues facing today's forests—I knew that every character was going to begin his or her story with the line *I love trees*. And when I showed people the script, there was only one recurring criticism: "Does everyone really have to say, 'I love trees'?"

My readers worried that it was too implausible, too stagy, and they couldn't wrap their heads around the idea that someone could love trees *and* cut them down.

But the line was non-negotiable.

I know that the people who cut trees down love them every bit as much as the people who lay down in front of bulldozers to stop them. That's why I find the battles over forests so fascinating.

Recently I was on a plane from Prince George, BC, to Vancouver, and I found myself sitting next to a woman who works for the BC Forest Safety Council. MaryAnne Arcand is known in the industry as "bulldozer" because of her relentless pursuit of higher safety standards for logging truck drivers. Arcand had read about *The Green Chain* and was wary of it. She suspected there was one thing I didn't realize when I wrote the script and wanted to set me straight. "You know that all these men love the bush," she said. "I've seen them cry at the sight of a beautiful tree falling. I've seen them offer thanks, the way the natives used to. They really love the bush."

The first interview I did for this book was with Wade Fisher, a man who started working in the forests about the same time I turned up on the planet. As we were talking I decided I had to ask him the same question I'd asked every character in the movie: "How do you feel about trees?"

There was a pause before he answered, "I think I love trees."

I could hardly breathe.

Since that interview I've asked that question more than two dozen times and roughly half the people have responded with those exact words. And very few of them had seen the movie before we met.

Before I started filming *The Green Chain*, I met the former chair of the Forest Stewardship Council of Canada, John Wiggers, and he explained that BC was "ground zero" for forestry issues because it's a place where almost everyone is passionate about trees. He said there was nowhere else that the convictions—and the divisions— ran so deep. And our battles are ongoing and iconic and have echoed around the globe.

Environmentalism is a British Columbian's birthright. Our province launched Greenpeace, the Sea Shepherd Society, the Raging Grannies, *Adbusters*, ForestEthics and Dr. David Suzuki.

But cutting trees is a BC birthright too. Our economy has historically been fuelled by forests. Our communities have been built around mills.

This is a place where people live and die for trees. And I'm not just talking about protesters willing to go to jail or live on platforms in the treetops. There's a reason BC appointed a "forestry coroner." Cutting trees is one of the most dangerous jobs in the world. If you ever doubt that, try driving on a logging road—and if you'd like to come back alive, I suggest you drive very, very, very carefully.

Besides asking people for their feelings about trees, I also asked for their solutions to the problems facing our forests. So this book is not only a collection of people's feelings about forests, but their best ideas for keeping our forests—and the forest industry—alive and thriving.

Not to spoil the twist ending, but most people I spoke to were not big fans of exporting raw logs, and almost everyone's very nervous about climate change. They're also pretty concerned about protecting our remaining stands of old-growth, or ancient, forests.

For so many of the people I met, the battle over the Clayoquot Sound rainforest was life-changing. When British Columbians talk about "the war in the woods," this was truly the "Great War" that put the fights over BC forests on the international map. In 1993, almost 1,000 people were arrested at Clayoquot over three months for acts of civil disobediences for their participation in the logging blockades. In January 2000, the region was designated as a biosphere reserve by the United Nations Educational, Scientific and Cultural Organization (UNESCO). But as I write this, disputes over which areas are being logged and which are being reserved for our biosphere are ongoing. There are times when it seems like the only natural thing being preserved for future generations are the rocks.

The other great battle of BC was fought over "the midcoast timber supply area"—52,000 square kilometres of wilderness that were saved not just by environmental protests, but by brilliant media spin when the protesters renamed the region the "Great Bear Rainforest." At the time I joked that perhaps they could save every tree in BC by renaming the spruce, the cedar and the pine Flopsy, Mopsy and Cottontail. Again, it's a battle that environmentalists thought they won, even though the terms of the agreement keep shifting.

When people around the world talk about the devastating impact of clear-cut logging, the image that comes to mind is of a site near Bowron Lake in northern BC—a swath of moonscape so huge that in the late 1980s it became infamously known as the only man-made object visible from space, other than the Great Wall of China—a chilling claim verified by the Canada Centre for Remote Sensing in 1991.

Almost everyone I interviewed was in favour of taking a longer-view approach to managing our forests. But Squamish Chief Bill Williams suggested that planning for the future doesn't mean planning for 50 years from now, it means planning for the world you want in 500 years. Velcrow Ripper echoed his sentiments, saying that we have to be looking forward seven generations. Not an easy concept to grasp for politicians trying to woo voters in time for the next election cycle.

Perhaps the biggest surprise for me was that I had braced myself to be depressed. Seriously depressed. And while it's hard to listen to tales of mountain pine beetles devastating more than 10 million hectares of forests (that's more than three times the size of Vancouver Island) without crying, then praying for a very long, very cold winter or three, most of the interviewees were surprisingly optimistic. They think the solutions are out there, and now that we're living in the age of Al Gore and green is the new black, our society might be willing to embrace the solutions. Or at least attempt them. So instead of becoming depressed, I came away inspired.

It's my hope these interviews will inspire you to conserve, respect and treasure our resources and do what you can to save the planet. Maybe you'll come away with your own solutions to save our forests.

I know they'll introduce you to some truly amazing people—all of whom really do love trees.

May 2009
Vancouver

1
GEORGE BOWERING
LOGGER LAUREATE

George Bowering is a poet, novelist, editor, educator, historian, critic and national treasure.

Bowering has published more than 80 books and has won most of Canada's major writing honours—including Governor General's Awards for both poetry and fiction, the bpNichol Chapbook Award— twice—and the Canadian Authors Association Award for Poetry. In October 2002, Bowering became Canada's first Parliamentary Poet Laureate.

So with such a distinguished writing career, what else would anybody want to talk to him about besides his work in the BC Forest Service in the 1950s—something he'd never been interviewed about.

Bowering's connections to the forest run deeper than his first job. When we spoke, his "kid brother" had just retired after more than 30 years of sharpening knives for a mill in Okanagan Falls, BC. And that wasn't the only family member connected to the forests. "My late wife's brother-in-law, he's a scaler."

Bowering was born in Penticton, BC, in 1935 and grew up in nearby Oliver. After spending three years in the Air Force, he returned to Canada and spent a year in university before dropping out temporarily ("It wasn't where I belonged at that time") to join the BC Forest Service. I met Bowering in October 2007 at his home on the west side of Vancouver to talk trees.

Tell me about your time in the BC Forest Service.

I joined the BC Forest Service in 1958—you know, when I was four. I was hired as a marker. I went to an area that later on became the subject of a couple of my novels: south of Kamloops, around Merritt and Ashcroft. It was basically the Merritt forest region. I lived in a bunkhouse in Merritt and I was on a crew with two other guys. Jeremy Crow was a great big, huge, heavy Englishman—with a German dog who used to eat raw hamburger—who beat the living Jesus out of me one time when we all got drunk. So much so that we broke the stove. And another guy, I can't remember his name, except he was the kind of guy you would expect would be a Steve.

This was in the interior of British Columbia in the late 1950s. And we had heard about a thing called clear-cutting, which some Americans did in California or somewhere. We thought, "What a bizarre, savage animal thing to do—no, not even an animal thing, a horrible human thing to do." To cut down all the trees in some areas! I can't believe it.

So the marker's job was to walk around the hills, the bush, with a big can of yellow paint—sometimes blue, sometimes red, sometimes white—but usually yellow paint on our backs, with a hose that came from the can and a [spray] gun in our hands. We'd walk around and mark the trees that the logging companies weren't allowed to cut. First, you would mark right around where the tree meets the ground at the bottom, so if there's a stump you'd still be able to see that. Then mark an X or something like that on the tree about eye level. So you mark the tree. In the wintertime, it got a little bit dicey because you'd be walking around with snowshoes and you'd mark this thing on the edge of the snow, and the snow would disappear and you'd find out you had another four feet of tree.

We'd mark the trees that we thought would be the best parents for the next batch of trees that were to grow. So if it was a tree that was a School Marm tree—you know what a School Marm is?

No.

It starts off with one trunk and breaks off to two trunks and goes up like a big letter Y. They can cut those. You don't want them all that much because they will leave cones that will grow more School Marms. You look for the best trees you've got in that sale—a sale is an area where they had bought the rights to cut in that area—and the trees that were at the tops of the hill, because if the tree's at the crest of the hill that means that the cones would fall from the tree, then roll down a certain distance and regrow that whole area.

So it was a lot different attitude—and one that has a completely different sense of time—than clear-cutting and then sending out a bunch of dope-using kids to plant trees. Because when those kids go in to plant trees, you get an area that's however many hectares of all the same tree. The trees all look the same and they're all the same height and they're all the same variety and they don't have that citizenship that goes with various other kinds of trees and various other kinds of bushes and so forth growing in that area, which you will get if you do it the way we said you're supposed do it.

Then another job—I was allowed to do another job when they needed somebody to do the most unpleasant part of it—I forget what it's called. It's the guys who go out and survey once the outfit, the logging killers, the tree killers, have the rights to cut these trees, you go in and tell them how much wood there is in that lot. You measure the whole sale and check the kind of trees there are, all the different varieties, and you tell them when they go in how many million board-feet of lumber there are in that sale [a board-foot is 144 cubic inches, or 2,360 cubic centimetres, based on 1 square foot of 1-inch-thick board]. It involves a couple of guys—one guy at the front of this long, long, long chain and a guy at the back of this long, long, long chain, surveying the outlines of this place. I was always the guy at the front of the chain and you had to go straight. If you came to a little stone canyon, you couldn't go around it. You had to go through it and then you went in and did all these calculations.

Well, after I had been in the service for a couple of months, 1958 became—up till then—the worst-ever forest fire season in Canada, in both BC and Quebec. The smoke from the Quebec forest fires, for instance, travelled as far as northern Europe.

In BC, we had one fire that was way up on the Yukon border, and the last I'd heard it had done like a million acres [405,000 hectares]. They said they wouldn't fight it. They would just wait and let the winter put it out because you couldn't get in there to fight it and it was too big to fight anyways. So you just had to let an area the size of Utah or something burn down. In Merritt, we had so many of these fires.

Cruising! That's the other thing you do. You go cruising, and then you go marking. I was cruising for the BC government!

▲ **Nice. I like that.**
So we couldn't afford to have anybody doing that anymore. Everybody had to be involved in the firefighting. Nowadays there are professional firefighters. They import them from Saskatchewan or Ontario to fight fires in BC. These guys make lots of money, big salaries. In those days, you had to get volunteers. They got paid hardly anything, just a little above zero, except they got a lot of free food, but they had to fight fires for like 17 hours a day. So here's how you got them to "volunteer." You drove a forest fire van around to the back door of the pub and then the rangers in full uniform went in the front door of the pub. The guys in the pub knew that you were a suppression crew, that they were going to get "volunteered," so they'd run like crazy for the back door

and the windows and the biggest guys in the forest service would be there to grab them and throw them in the van.

▲ **So these were serious volunteers then, these were dedicated—**
Yeah, absolutely. All they wanted to do was fight fires and not drink beer. The way it worked was that the fire boss guy, the warden or whatever, he had the power to arrest, like a cop does. So that's how they could get around this when somebody said, "Wait, this is kidnapping, isn't it?" "Nope, it's not kidnapping, I've got the right to, etcetera."

The job I had was lovely. All our vehicles were in use, there were fires all over the place and there were suppression camps up there fighting these fires. And on the top of just about every mountain there was a fire going on. So we had to rent vehicles from anybody who had vehicles that could do the mountains.

The vehicle I was assigned was beautiful. It was the smallest Land Rover you could get. It was bright yellow and it had pictures of chainsaws on all sides and the roof because it belonged to the McCullough chainsaw company.

My job was to go every day to the Safeway or the SuperValu or whatever the heck it was and load up. I had a big shopping list that I got the day before from the fire crew I was supplying—and I'd go in and buy all this stuff and load the Land Rover up and get the most recent Vancouver newspaper, which was probably published the night before, and drive halfway to the fire camp, go off into the bush for a little while with the paper and have a bottle of pop and so on and so forth, and then give the fire crew all the stuff I bought. Then they would give me the list for the next day and off I would go.

I was not a very experienced driver, I was just a cocky young fellow and it was really neat. You go up a mountain road, a logging road, and if a logging truck was coming down that road—the road was only wide enough for the logging truck—you had to shove your vehicle into reverse and go as fast as you could because the logging truck is coming down with its brakes on and water coming out underneath the truck and so forth. You had to go as fast as you could and find a place to turn around.

▲ **I know how small those logging truck roads are. How did you not end up killing yourself?**
I don't know. But as you're driving up this road, you're looking all the time for possible places to get off the road. So you know where they are as you're coming back, and you choose among them—like the one that's over the chasm or the one that's over a hill on the other side of you. That was a wonderful job.

▲ **Do you remember when clear-cutting came to BC?**
I remember being astonished when I saw it actually happening. I think I must have heard about it in the 1960s. I didn't see it, except in the States. I saw it in California, I saw it in Oregon, I saw it in Washington. I thought, "Oh my god, what are they doing?" I couldn't believe it.

There are clear-cut people, logging people, who are interested in the bottom line, who say that that's the most economic way of "harvesting" logs. That any other way you do it is not as time thoughtful.

When you're around a place like Vancouver, or if you're on the side of a heavily used forest road, you don't see as many clear-cuts. But if you fly over them, they can't hide them from you. They can't tell the pilot not to fly over the clear-cuts. But they are clearly ashamed of them, or feel as if they are doing something "criminal."

▲ **Well, we do a nice job of making sure the fringes of the highways look really good in this province.**
The first time I saw that was driving through the redwood forest

in California—this gigantic cathedral of trees. I had to go and have a leak. I pull off to the side of the road, go in among the trees and they're not there anymore. There's just a couple rows of trees on both sides of the highway and then no trees.

So it's like a Hollywood facade.

Yeah. You know what it looks after they've done clear-cut logging? You see all this scrap. They don't clean up. They just say, "Oh well, in 7,000 years, it will dissolve. It will somehow rot and become part of the eco-culture." It's hideous and awful.

Then I began to see it in BC.

There is something, I don't know what it does to the human heart. If you grew up in Oliver like I did, there is a great big, huge open cut on the face where they're getting out this white stone that they like for some reason—silica—they've been cutting silica there for decades. It's just hideous. When you get to the south Okanagan and you see all the dry brown and those few scant trees and the cactus, it's the landscape that the creator of everything finally came to and understood was like what he meant all along. That was like *the* landscape.

So if you come and just blast open the side of the hill—that is a sacrilegious thing to do. Somehow or another you know when you're looking at it that it's wrong. There is a relationship between beauty and good. There's got to be.

So when you see clear-cuts, it's that way too. That's why people hide them. I think maybe one of the reasons they hide them is they have friends in the tourist industry, and the tourist industry says that they'd lose some tourist dollars if visitors had to drive along clear-cuts.

How do you feel about trees?

I hug them every chance I get.

Two doors to the left of us is this house they ripped down

last week. And they put these little protective things around the trees that belong to the city that live outside the sidewalk. But the trees that were on the property, they just ripped them down—tore them up! There are chunks of them lying around. And I think I'm a pretty typical BC person because it just bothers me, it just hurts me when I see something like that happen.

This is the only place I know where there are headlines if somebody gets in trouble for cutting down a tree. Remember there was that story last year about that woman who wanted her view to be better in English Bay, so she poisoned the trees that belonged to the city across the street from her. She slowly poisoned them.

▲ **And then she had to go into hiding.**
Yeah, because people were really mad.

One of the biggest new books from the last couple years was that one about the sacred albino tree up there in northern BC.

▲ *The Golden Spruce.*
No one knows exactly where that tree is. But boy, in Canadian terms that was like a bestseller book and much rewarded. I'm a typical BC person that way. When I was a kid, I used to just think how heroic those guys looked in the ancient pictures of people killing trees. If you look back at pictures from, say, the 1890s, 1910s, you see pictures of guys standing there, all over BC in all the little towns like Kamloops or wherever it is, and you see them standing there over this desolate landscape. There's a picture of New Westminster when it was first de-treed, and you could see these people standing over the fallen trees like they were their victors. Then I realized they're similar to another picture. They reminded me of the picture in my Bible of Goliath lying on the ground and David standing with one foot over his chest brandishing his little slingshot, saying, "I took care of this

guy." So you'd see that picture of the g
over the logs. And I just hear that expr
or he had a fight against nature," or
    Then you would also see pictur
BC artist and poet—his dad was f
whaling business in Prince Ruper
guys who have conquered a whale. It s ᴗ
Instead of dead trees lying on the ground, it's a aᴗᴗ
on the ground with these human beings standing over uᴗ
saying, "We did it! We've reached some acme of human possibility." We knocked these trees down.

You get these images of these guys with these gigantic
crosscut saws with one guy at each end putting their saw in a
tree that was unfortunate enough to be the size of a house, and
they've sawed the thing and they're standing there looking like
they're heroic gentlemen. It's awful. It's like the guys fighting the
dragon. What was the dragon? Why was it so important to show
that a human being could defeat a dragon, or a tree, or a whale
or a grizzly bear?

**If somebody put you in charge of the BC forests tomorrow,
what would you do?**
I'd say, "Let's have a hiatus." My friend David Cull, he used to be
a poet. Now he lives in a little hillside town in Italy, but he's been
all over the world finding ways to build houses where you don't
have to knock down a bunch of trees. He built a house in the
Kootenays out of rubber tires—he used rubber tires!

If I was put in charge, I would designate authority here and
there. But if you had to have a bunch of people running the
forest region of, say, the northeast, I'd put in a forester and a
poet, and they'd have to work it out together.

George Bowering at www.library.utoronto.ca/canpoetry/bowering

2

TOM McPHEE

MILL WORKER

Between the summer of 1985 and the spring of 1986, I worked as a reporter for the *Williams Lake Tribune* in Williams Lake, BC, and the staff photographer, Wendy Holm, quickly became one of my best friends. In 1993, Wendy married Tom McPhee, a local mill worker. Tom and I have been friends ever since.

When I was offered the chance to turn my script for *The Green Chain* into a movie, I asked Tom to read it for me. I decided that if Tom didn't think the stories felt real, there was no point in making the movie. After Tom called and said, "I know exactly how the logger is standing" and went on to describe perfectly and precisely the image of the character I had in my mind before he added, "That's my neighbour," I phoned my producers and told them we were ready to roll.

Tom is someone with an opinion or two about pretty much everything. He's a soft-spoken, no-nonsense mill worker who moved to Williams Lake as a teenager in 1972. Unlike the lumberjack stereotype, Tom's not a big, burly guy. He's in good shape, but good shape at 5 feet 7 inches, aged 48.

Like so many forestry workers, Tom dropped out of high school to work in the mill. His mill was Lignum's—and for most of his life Tom has been turning trees into *dimension lumber*, a term that meant nothing to me until he translated it as "two-by-fours."

In 2006, Lignum's was taken over by a company called Riverside. Tolko Industries—a privately owned forest company based in Vernon, BC—purchased Riverside the following year.

In 2008, the Williams Lake lumber mill closed "temporarily" and Tom was given the "opportunity" to retrain.

Not everyone in Williams Lake works in the mills, but when I was living there it sure seemed that way. I was in my early 20s, so it was hard not to notice that every other guy my age seemed to be driving a shiny new truck and all the newly married young women—and every woman over 18 seemed to be married—had substantial rocks on their wedding bands. It was clear to me that mills equalled money.

One of my first assignments as a reporter was profiling the local mills. At that time, there were five major mills and several small ones. Today, two companies own every remaining mill in town, and with those remaining mills downsizing and regularly decreasing shifts, mills no longer equal money.

But it still seems like most people in Williams Lake either make their living at the mills or are connected to someone who does. Tom's brother-in-law Ken worked with him at the same mill, spending 36 years there, then retiring just before the layoffs. One of Ken's three sons started working at Tolko six years ago. His two other sons work in other mills.

Three weeks after the mill reopened—in May 2009—Tom and I talked about vanishing mills, vanishing jobs, vanishing expectations and the toughest job he has ever had—working on the green chain.

▲ **How long has forestry been part of your life?**
I didn't go to grade 12, so I started at the mill at 18. And this June 19 [2009], I'll finish 30 years, start 31 years.

▲ **What was your first job?**
Repiling lumber out in the yard. And then I moved on to the green chain.

▲ **What's a green chain?**
After the lumber has been planed, it moves out from the planer, through the trim saws and then it comes out on the chain once it

has been cut to length. And based on the length and the grade of the boards, you pull the boards off into the correct piles. There are five chains, and you determine the length of the board by which chain the board ends up on. It was the hardest job. I was only 147 pounds at the time. I don't know how much a 20-foot two-by-ten green fir weighs, but I do know it weighs more than I did.

▲ So how did you do it?
The boards are on rollers, but to get [the rollers] moving, you have to pull. The bigger guys could just grab them and kind of pull, but I had to brace my foot up against the wall and give it a pull.

▲ How do you figure out what the grade is?
Basically, they show you once and expect you to learn it. [Laughs.] Back then, there were only four or five grades. Now we have up to 10 or 12 at any one time. But the one and twos are stamped and the number threes are stamped. And the economy and the rejects—or if we had a special one—are stamped. So we could go by the stamp. And, of course, the length. There's like five chains, I believe, that they run on rollers, and you determined where the lumber fell on what chain and you knew what the length was and pulled it off into your pile.

▲ Why was that the hardest job?
Just the physical aspect of it. Like I say, I'm pulling lumber off the chain that was heavier than I was. And by the time you got done, the piles were taller than I was.

▲ What are all the other jobs you've done?
I've done just about every job at the planer because I was charge hand for 15 years, and when you relieve people to go to the washroom or whatever, you have to be able to do their job. I've driven every piece of equipment on the property, right from the

little warehouse forklift to the Wagner logstacker—which is the largest machine there [a heavy-lift machine used to load and unload logs from trucks]. And I did various jobs at the sawmill. I've run the barkers [which remove the bark] and trim saws, but the key ones I haven't done are the chip-n-saws and the head rig [which cuts the biggest logs into lumber].

▲ What's your current title?
Since the layoffs, I've been back to loader-operator.

▲ And before the layoffs?
I was the crane operator, a picker—the one who stacks the logs 35 feet [10 metres] up for storage.

▲ One of the things I found fascinating was that your layoffs weren't called layoffs—there was some wonderful euphemism they used.
We were put on the "work share" program. So we were on for a four-day workweek for a month, and then we were on a three-day workweek for one week. The next Monday, they actually said, "Everything's going fine, we expect everything will be great." And that Wednesday they told us we're laid off.

▲ How long did the layoffs last?
I was laid off for, I think, two months. Then they called us back three weeks ago.

▲ Do you know how long you're back for?
No. They said as long as we don't have a rush of accidents and our production stays up to, I think it's just over a million board-feet a day [a board-foot is 144 cubic inches, or 2,360 cubic centimetres, based on 1 square foot of 1-inch-thick board], we should keep running because we are the most efficient mill in the Tolko fleet. I believe

there are 16 various types of mills and a couple of plywood plants and things like that, but we're apparently the most efficient.

> **What kind of wood are you working with? Are you working with old-growth or second-growth?**
> I believe most of it is second- or third-growth. And right now we're mostly working with bug kill [trees infested by the mountain pine beetle].

> **What do you have to do to deal with the beetle-kill wood?**
> Be very gentle with it to get it in the planer. Normally we can drop a log. Now if we drop a log it will usually break into three pieces. So we have to get it into the mill and get it cut. And then it's just as strong as dried lumber because we cut all the defects out of it.

> **You decided to do some retraining when you were on your "break." Can you talk about how that happened?**
> When we were laid off, we found out there was some money from the provincial and federal governments for retraining. We heard rumours that the mill was going to start up again in September or October. We didn't think we were going to go back that early. So I was trying to think of a job I could still do even when I went back to the mill. I was thinking about taking a welding class, but that's six months and it didn't start until September. If the mill started again in September or October, I would have had to make a choice—either quit the mill and continue on with the course or quit the course and go back to the mill.
>
> So I took my Class 1 driver's training. I can drive semi-trucks, logging trucks, bus, school bus, city bus, taxi, it's all in there.

> **You've managed to keep working all this time, but how are other people you know doing?**
> People—you can tell they're angry. They get upset really quick

over nothing. And they don't know what to do. Myself, I know I was quite concerned before I got called back. I've worked at the mill for 30 years. I've put everything into making sure our house is paid as quick as we can. We had no bills or anything, but if we'd been off for a great length of time, I could have ended up losing everything and there's nothing I could do about it.

▲ **Who are people mad at?**
To me, it seems like they're mad, and they really don't know who to be mad at. They're going to lose everything, and they can't do anything about it. They can't just go find another job. You get 600, 800 people out of work in the same town, there's just no jobs.

▲ **So are people leaving town?**
Some people have left town. A lot of them are tied here. They have 10, 15 years into their house, and they just have a few years to pay it off, so they were just hoping the mill goes back. They're angry that they're going to lose it, but they've got no one to blame, which makes them frustrated. You just can't do anything. And where are you going to go? A lot of other small towns in BC are forestry-minded. They're all shut down. Mackenzie [a logging community in northern BC] shut down because the government has let the lumber companies ship their logs anywhere they want [wherever it's cheaper to process the wood]. So we're getting logs from Mackenzie . . .

Our mines—we've got two mines here—they've each laid off approximately 80 guys. So you can't go to the mines. Some of our millwrights did get jobs at the mines because they were short of tradespeople. But a lot of us aren't tradespeople.

▲ **I remember when I was there, whenever anybody lost a job it was the environmentalists' fault. That was who got blamed.**

There's no sense of that now?

Not really. With the different rules now, we can ship logs from any-where to get them . . . You can make as much lumber as you want, for as cheap as you want, there's just nobody out there to buy it.

▲ **What else should I know?**

People are a little scared, and they just don't know what to do. Like myself, what do you do? You can get retrained, but where are you going to go? I had a hard time figuring out what I wanted to be trained for. I'm 48 years old, what do I get retrained as? I unfortunately never finished my grade 12. I've got a lot of interests, but what can I make a living at?

▲ **It was always my sense that grade 12 had no real point in Williams Lake because so many people were able to walk into really well-paying jobs.**

Oh yeah. I was at the mill and it was a great job. Still, young kids don't finish school and come to the mill. But a lot of them I've been seeing, especially over the last five years, come to the mill and say, "This isn't what I'm going to do. I'm going to go do something more enjoyable." When you're doing a job, you're more than likely going to be there for a very long time. Thirty years to me is a very long time, and I really don't enjoy what I'm doing, but I'm stuck. Where can I go?

▲ **Did you have a favourite job at the mill?**

My favourite job was driving the Wagner, the biggest machine we own. I just enjoyed it, that feeling of driving something that big.

▲ **If somebody put you in charge of our forests, what would you want to do?**

I'd like to bring back the appurtenancy clause [a BC government regulation removed from the Forest Act by the provincial Liberal government in 2003], which says if you cut down trees in this area, they must be milled in this area. I mean, other than that, we're selling our jobs. I don't know what they expect us to do. If you sell the logs [out of the community], what are we supposed to do for work? I just don't understand that thinking.

▲ **How do you feel about trees?**
Trees have been the biggest part of my life. They've put my kids in college, they gave me a nice home, vehicles, everything. Like I say, the government's selling them and if we don't get to mill them, I don't have a job. So trees are a very big, important part of my life.

3
BETTY KRAWCZYK
FANATIC

Betty Krawczyk is proud to call herself a "fanatic." The infamous Krawczyk has been in and out of jail so often that she should be collecting frequent flier miles. At the very least, with every 10th visit she should get a free pack of cigarettes and a harmonica. Krawczyk did her first jail stint in 1993. She was 65 years old. Her crime: blockading logging trucks.

Since then she's spent more than three and a half years in three different jails serving eight different sentences, and she has no plans to slow down. She's been called "Vancouver's best known raging granny"—and she's definitely raging and a granny—although she's only an honorary member of the satirical seniors protest troupe that was born in Victoria, BC. While the Raging Grannies are a team, Krawczyk tends to be a one-woman band and a one-woman brand.

She credits her birth in southern Louisiana for her passion. "All southern Louisiana people are too passionate for their own good." Her first protest—an antisegregation rally in the early 1960s—changed her life. "I joined a small group of white people called SOS, Save Our Schools. We went down to the same elementary school my children were going to and we picketed the school with signs that said, DON'T CLOSE, INTEGRATE. LET'S BE CIVILIZED, INTEGRATE OUR SCHOOLS. DON'T CLOSE. CLOSING IS DEFEAT. There were only seven of us. It was the first time I'd ever been spit on."

Thanks to segregation, Krawczyk split with her church—because they sat out the fight. She went on to protest the Vietnam War before moving to Canada and discovering her next cause, saving the forests around her new home, Clayoquot Sound.

I interviewed Krawczyk in May 2008 at her apartment in Vancouver's east end—a place packed with boxes full of papers from dozens of battles, including running for mayor of Vancouver with the Work Less Party.

There's an eclectic collection of art on her walls—and I'm sure every piece has a story behind it and the memory of a fight or two—but what stood out for me about the decor was all the art by and about children. This is very much the apartment of a grandmother—she had eight children and has eight grandchildren and one great-grandchild. And she's a grandmother determined to spend the rest of her life fighting for a better world for her grandchildren and everyone else's.

Krawczyk regularly shares her adventures, experiences and opinions on her blog at www.bettyk.org, and she has written three memoirs: *Clayoquot: The Sound of My Heart; Lock Me up or Let Me Go: The Protests, Arrest and Trial of an Environmental Activist and Grandmother;* and *Open Living Confidential: From Inside the Joint.* Her latest book is entitled *Are You Crazy, Lady?*

I talked to Krawczyk about her adventures in activism, life in prison, why the world needs more fanatics and her accidental discovery of environmentalism after she moved to BC and all her children had left home.

▲ **How did you get involved in the environmental movement?**
I bought 10 acres [8 hectares] in the Clayoquot Sound on a place called Cypress Bay, and one of my sons built an A-frame for me. It was the most wonderful experience. I lived out there for three years.

It struck me through the heart that the beauty of the place was being demolished by the way the logging companies were just absolutely clear-cutting everything in sight. And as I went back and navigated through the Sound to see some of these more remote places where the clear-cutting was just crazy, you could see that whole mountainsides were sort of washing down into the sea from soil disintegration.

The 10 acres that I had bought was old-growth. There had been some selective logging through there, but it was primarily old-growth. As I became acquainted with the huge mountain behind the place, I saw that what looked green from a distance was up close not really trees, it's shrubs, it's scrub brush. And during the winters when the heavy rains came, that scrub brush didn't hold that mountainside in place, it all slid down. And part of that was coming down around my place.

I got up one morning and went out and there was this humungous landslide sitting on the beach on the cove, covering almost all the cove, certainly covering the creek where I got my water, the stream, and I looked at that and thought, "This is outrageous, this is absolutely outrageous, they're just killing the earth with this clear-cutting."

The debris was washing into the streams. These are, or were, fish-bearing streams that were just being destroyed by the greed of logging companies. So I went to Tofino and complained to the forestry and the fisheries ministries—which did no good. No response. Then I went down to Victoria and complained; still no response.

There was a group in Tofino called the Friends of Clayoquot Sound, and I had been stopping in and giving them bits of money from time to time. But now I knew it was going to take more than that; that all the issues I had worked on before and written about and thought about and researched, they were all predicated on a reasonably healthy Earth. If you don't have a healthy Earth, you don't have anything. Everything else, it's as though it might not even exist. Without the Earth being healthy, nothing is healthy, nothing will live. So I became convinced that all these other issues had their origin in the way the Earth was thought of, in the way the Earth was treated. I made a commitment at that point to the environment as my friend—a total commitment.

The night before I went out to the blockade, in the summer of 1993, I stayed up all night making sure that I would be able, psychologically, to honour this commitment, whatever it may bring, because I'd never been arrested before and I was fearful of being claustrophobic. And I worried about that. I worried about making an ass out of myself. What if I was the only one being arrested? I didn't know if anybody would be arrested because it was the first day of the blockade.

So I stayed up all night, I didn't sleep at all, and by the morning's light I had made peace with it that even if I made a fool of myself or I was imprisoned, that was okay.

Were you the oldest person there?
Oh yes, I'm always the oldest person anywhere. [Laughs.] I'll be 80 in August [2008], you know.

So I found that I could stay in jail just fine. I have an active inner life that I can retreat to when the outside becomes too raucous on the nervous system. As a writer, you know how that goes, you can retreat to your own places, imaginary or otherwise. [Betty noted that two other women, one of whom may have been older than her, were also at the blockades.]

How did the other protesters treat you, and how did the police treat you the first day you were arrested?
They were good. Cameras were on them. No police officer wants to be accused of throwing around the fragile bones of grandmas, so they were careful and it was fine.

And then three of us refused to sign the undertaking to say that we were sorry or that we would be good, so we were sent to the men's prison in Nanaimo, Brannen Lake, in a special unit. Thus began our prison terms. But looking back, that actually was very soft treatment compared to what I've endured in women's prisons on long terms.

▲ **Really?**
Yes, because there were so many of us that were arrested in the Clayoquot Sound, it was a new thing. Because so many of us were vegetarians, they brought over boxes of vegetables and oils and stuff and let people cook their own stuff. It was pretty easy. And we were allowed to talk to the press.

▲ **The impression that I got from watching you on the news and reading various stories about you is that, if anything, prison strengthened your resolve.**
It did and it does still . . . It does strengthen my resolve because what happens is that I've been through the worst that they can do to me and survived it, and I'm not afraid of it. And when you're not afraid of the worst that they can do to you, then you have a power that they don't like, that rattles them.

▲ **You wrote on your blog recently about being a fanatic. Can you talk about being a fanatic?**
[Laughs.] Well, I am a fanatic. But we have to become fanatics in order to make changes. I can go down and vote right now, for whatever that means, because of the fanaticism of earlier women. We can join a union if we want to because of fanaticism of men and women, the Wobblies [Industrial Workers of the World union] . . . There was a huge movement of fanatics that came together and said, "We want more life." They were fanatics of life. What we need is fanatics.

▲ **How do you think the Clayoquot blockade changed things?**
For one thing, the Sound became a biosphere reserve. But that's

actually not the main thing. They're looking now at wanting to go into one of the pristine watersheds and log. The First Nations people there are quite on side with logging companies and have always been. A lot of First Nations people are sucked into these agreements, and they're poor, they have been poor, so they will take what the logging companies have to give. I don't want to comment on what I think of them.

But nevertheless it's not finished yet in the Clayoquot Sound. But one thing it did do was bring to worldwide attention the determination of these logging companies to cut down everything they could possibly cut down.

I can remember arguing with loggers in the Elaho Valley, when we were blockading there, about Interfor. I remember this logger saying, "Why do you keep saying Interfor does this and Interfor does that? We are Interfor!" Well, now they know that they aren't Interfor. Interfor cuts and runs like they've always done.

And now they don't even have their union. The IWA [Industrial, Wood and Allied Workers of Canada] isn't even there anymore because the logging companies managed to kill their union too, along with help from the government . . .

They've got to get smarter than that. People have to get smarter than what they have exhibited in the past. We all have to learn from our mistakes. And I've certainly had to learn from mine.

The union just went down the drain because there were no longer enough people working in forestry to keep it going. And it wasn't from anything the environmentalists did or didn't do. It was from mechanization in the forest and the mills, and a provincial government that doesn't care diddly-squat about the people of this province. What they care about is their ideology, of the only way to run anything is by privatizing everything. They are determined to privatize every public natural asset in this province. They're damming our rivers as we speak here.

So the Clayoquot Sound was a pretty thing, in a sense. Did it change anything? No. Except for the worse. It changed everything for the worse because then people think everything is okay. They think there's going to be a biosphere and things are settling down in British Columbia.

Things are not settling down in British Columbia. They're cutting all over the bloody place. What they're doing is taking public land with trees on it, with forests, declaring it private land, cutting the trees and then selling the land to developers. This is what's happening.

I tried to tell these loggers years ago, "Look, if you guys would petition government and tell the government you're going to do this, and my god stick by it and stand on a blockade—you've got a union, say you want community logging practices, that you want to start having communities in charge of tracts of forests and do selective logging."

But no, they're brainwashed into thinking that there will be no logging unless some huge multinational, preferably American, company comes and does it and hires them. Otherwise they have no notion of trying to take possession of a resource that is theirs to start with—it belongs to all of us—and petition for the right to manage it in a selective way. Everybody would still be working in the forest if it was selective logging.

How do you feel about trees?

I've always loved trees. I was raised in the swamp in southern Louisiana with cypress trees and oak trees and what they call a "live oak tree" that stays green year-round. And as children we played in the wooded and swampy areas. So I grew up feeling kin to the natural world.

But primarily trees for me are families. When I see an individual tree standing alone, I always feel sorry for it because it's separated from its family. This is why forests present such an enormous thrill to people when they go into a real forest. They don't exactly know the composition of the thrill that they feel. It's even in the breathing, the air of a forest is very different—but it's because you're in the midst of a thriving ecosystem composed of families.

When I give talks in the schools, I try to explain to the kids the difference between an old-growth forest and a tree farm by giving them something they can relate to. I say, "You've got a big area about three times the size of this school. In this forest, you have grandmothers and grandfathers, mommies and daddies, uncles and aunts. Then you have big brothers and sisters and little babies. When the wind blows, the mommies and daddies and the grandmas and grandpas keep the little ones from blowing away.

"To keep the little ones from being flooded away when it rains, the big ones collect water and release it gently to the little ones so they can drink. And then the grandmas and grandpas—because they have to be at least 100 years old before they can grow this lichen—have this lichen fall off them to the ground when the wind blows. The little ones eat it, like vitamins, and it makes their roots grow strong and stabilize.

"Now the difference between this huge family and a tree farm is that the trees don't just grow. The people come in and put down all these rows of trees that are all the same size, they're all the same age. And they've been altered chemically. They're not exactly natural. And they don't have any mommies or daddies. And people come periodically and they poison them with herbicides and pesticides and some of them die, a lot of them die. They're washed away. They're blown away."

And at that time the little kids are all saying, "Aww. Aww." And the teacher's getting alarmed.

⬛ **If somebody turned around tomorrow and said, "Betty, you're in charge of Canada's forests," what would you do?**
Immediately, I would call a meeting of all the activists that I have

known. Then I would call people whom I consider experts but that haven't necessarily been activists. I would call people like Tzeporah Berman and David Suzuki, Joe Foy and Ken Wu. But I would also call the people who have lived in the forests, who know the forest from different angles. I would call First Nations people, who have not been, what I consider, too eager to make deals with the government. And there are quite a few of them who have protested the destruction of the forests. I would call the ones who are considered really radical First Nations and I know quite a few of those.

We would all sit down together and not have one single government person there because governments have got us into this and are simply in bed with corporations. We will not have any corporations there. [Laughs.] This is going to be a people's forum because it's the people's forests. These forests belong to all of us, and they belong to First Nations, and they belong to people in general.

And then we will come together with the First Nations people and say, "Okay, we're not going to have any industrial logging in British Columbia—that is out. There can be a certain amount of selective logging. Now, what are we going to do here? What areas do each of you think would be able to be logged in a selective manner that has not been already too compromised?"

We will save the pristine valleys, but we have to figure out what to do with the beetle kill. This is a huge problem.

And let's get some real connection going here between activists and scientists and knowledgeable people who understand bio-systems but who also understand that we are going to have a different way of considering the resources of British Columbia and they will not primarily be economic. They will be economic in the sense that a living can be made out of them, but not obscene amounts of money by a few people who steal everything away from the rest of us.

Are you familiar with Jean-Paul Sartre's existentialism theories?

A little. I have to flash back to university to remember them. Well, they'll break your head, really, if you try to understand them. I've incorporated a couple of things he says into my own activism: that we are each unique, and there's nobody like us in the whole world, and everybody has within them, no matter how they've been conditioned, an element of choice. So with each person being so unique and having somewhat of an element of choice, each person is a representative of the human race. And therefore every individual decision is important because, in a sense, you're deciding for the whole human race. So I think of this on blockades. "Well, okay, I'm making this decision. I decide that this forest should not be cut down." Or I make a decision that we're going to have clean water out of this creek. Then you have the peace that comes with these decisions. That's what it promises. And that's the only thing the universe promises you. But it's huge—it's really huge! It enables you to live not a contented life, but a life that's free of unnecessary worries and strains and anxieties because you just feel more connected with the universal process. That's it. You feel more connected to the universal process.

It brings a lot of—I hesitate to say it, but sometimes I'm flooded with joy. [Laughs.] It seems crazy, considering the kind of work I do and the doomsday scenarios. But nevertheless, sometimes I just feel really happy that I got to live this life. And look, you go outside and see the green, it's beautiful, go down to the water. You know it's a great place. The Earth is a great place to have landed on. So we need these times of celebration too.

And that's it. That's my story, and I stick to it.

Betty Krawczyk at www.bettyk.org

4
KEN WU
OLD-GROWTH-TREE HUGGER

British Columbians care about ancient forests.

On October 25, 2008, the Western Canada Wilderness Committee (WCWC) organized a rally at the BC Legislature in Victoria, BC, and 2,700 people showed their support for saving BC's remaining stands of old-growth forest. The turnout tied the rally record set at the height of the protests to save Clayoquot Sound. And those 2,700 people were a mix of environmentalists and forestry workers.

No, that's not a typo.

The protesters included environmentalists *and* loggers working together, which to some people might once have conjured images of the great Dr. Peter Venkman in *Ghostbusters* fretting about end times— "Human sacrifice, dogs and cats living together . . . mass hysteria!"

The man who is arguably most responsible for the dogs and cats playing house is Ken Wu, the 35-year-old Victoria campaign director for the WCWC, or, as the group is lovingly known in the environmental movement, WC Squared.

Wu became environmentally aware when he was three, developing an affinity for frogs, toads and snakes while growing up in southern Ontario. At age nine, he discovered the nonfiction section at his school library and started reading everything he could about animals and exotic terrains like the African savannah. When his family moved to Saskatchewan that year, he fell in love with trees. "I just remember noticing trees and thinking, 'Those are actually pretty cool.' And I started getting the concept of wilderness and ecosystems." Wu's father was an activist for independence in Taiwan, "so the political aspect came naturally to me."

Wu started tree hugging with the WCWC at age 19 when he volunteered to stuff envelopes. That led to a job as a door-to-door canvasser raising money for the cause. Since then he's done everything he can think of to save old-growth forests, including showing up at one of his first rallies wearing a bear costume. In 1999, he was hired as the WCWC campaign director for Victoria.

I met Wu at the Legacy Gallery and Café in Victoria just a few days after his record-tying environmental rally to talk about how he had helped to assemble "the biggest active grassroots movement in the province," why saving forests isn't just for tree huggers anymore, how Facebook changed the face of activism and why "revolution is just around the corner."

△ **What first got my attention about you was the story about the loggers and environmentalists laying down the chainsaws and the picket signs to team up against the government. How did that happen?**

When we were blockading timber workers in the 1990s, when I was with other environmental groups, I had also been following what was going on in California, and I'd spent some time in Oregon and California, where they were actively working to make alliances with the mill workers in particular. Now they had a lot of difficulties doing that, but they did manage to make some alliances. And Judy Bari was a California activist who pioneered the whole thing in the late 1980s, early 1990s. I just basically followed what she was doing, but we didn't really put that into practice until the year 2000, when the Youbou sawmill on Vancouver Island near Cowichan Lake was being closed.

TimberWest was looking to shut down its Youbou sawmill, which would let go 220 workers. At the same time, TimberWest was exporting huge amounts of raw logs that could have gone through that sawmill. I thought, "Shutting down that sawmill doesn't mean the company's going to cut less forests. They're going to cut down all those trees that they normally would have, but they'll simply export those logs instead of creating jobs [in the community] We can take a stance that we want to see that sawmill continue to stay open because it's not going to affect the rate of cut if they shut down that sawmill."

So I spoke on behalf of the Wilderness Committee at a public forum with Ken James, one of the Youbou sawmill workers, and we both stood on the same stage opposing raw log exports and saying that we needed to keep that mill open. And the room—full of 60 or 70 mill workers—gave huge applause when I said, "We as the Wilderness Committee support keeping that mill open."

From there we showed up at their demonstrations and protests and then the Youbou Timberless Society leadership started showing up to speak at my rallies. And eventually they were bringing out more rank and file. And then it led into the unions— the Pulp and Paper Workers; the Communication, Energy & Paperworkers; and the Steelworkers—working together where we agree, which is on banning log exports and ensuring sustainable logging of second-growth.

▲ **Did anybody at the Wilderness Committee die of shock when you said you were going to stand up and fight to keep a mill open?**
[Laughs.] I think at that point people realized that it was a frivolous fight to spend your time fighting tens of thousands of workers. Not only that, it's counterproductive. If we could create a win-win situation, which is to save the old-growth, ensure sustainable logging of second-growth forests and ban raw log exports, then the workers will be happy, there is a long-term interest in maintaining a sustainable resource for forestry workers and at the same time we could protect old-growth forests, which are incredibly scarce now.

▲ Was it hard to get the loggers on board for not logging old-growth?

I wouldn't necessarily say that all the loggers are on board. But we work together where we agree. And most of the forestry workers, and certainly the mill workers, know that old-growth is coming to an end. There's not a lot of it left. The reason the coastal forest industry is ailing is not simply because of the high Canadian dollar, until recently, or the collapse of the US housing market. But fundamentally its long-term decline is driven by the depletion of the resource. They've taken the biggest and best trees in the valley bottoms, the accessible areas, and now they're only left with the scraggily trees in the far west of the island on steep slopes that are expensive to reach and yield lower returns. So they've got to shift to second-growth, and it's already happening. We're just saying, "Let's shift to second-growth, complete that transition, before we run out of old-growth." It's going to happen anyway. So let's do it now, rather than later, before it's all gone.

▲ Who came up with the ancient forest campaign?

In spring 2006, we got satellite photo data of how much old-growth forest remained on Vancouver Island. Prior to that we had primarily been campaigning valley by valley—to protect the Walbran Valley, the Nahmint Valley, the upper Sitka—but once we looked at the maps and did the statistical analysis of how much old-growth remained, we realized that for the most part Vancouver Island consists of tattered remnants, kind of like Swiss cheese, of ancient forests in a landscape dominated by clear-cuts, with the exception of Clayoquot Sound, which still has large blocks left. But at this point, it doesn't make sense to fight for one tattered valley after another tattered valley; let's get what remains of the old-growth protected because there is so little of it left. So we made the call at that time, about two and a half years ago, to protect the remaining old-growth on the island.

There's some nuance to that—the mills are still geared toward old-growth. So we're saying that they need to phase out old-growth logging across the entire island and lower mainland by 2015. That gives time for sawmills to retool for second-growth. That way we can maintain the jobs of forestry workers. So it would be a graduated phase out across the southern coast, so that within six years' time there will be no old-growth logging, less and less old-growth logging every year until it's done. Also, we do want to see an immediate ban in all flat areas, low-elevation valley bottoms, where 90 percent of the old-growth is gone; south of Port Alberni, where 88 percent of the old-growth is gone; and the eastern side of Vancouver Island, where 99 percent of the old-growth is gone. There's got to be immediate bans there. As well, an immediate ban on logging of all habitat needed for the survival and recovery of the spotted owl in the southwestern mainland. Other than those areas that would be immediate bans, there would be a graduated phase out over six years.

It's like Kyoto with climate change. No one expects that they're going to stop using fossil fuels tomorrow, which is why you have emissions reductions like 20 percent by 2011 and 30 percent by 2020 and so forth until it's fully phased out after several decades and it's all renewable energy by then. Similarly for old-growth logging; no government is going to stop it all tomorrow for the simple reason that there are still thousands of people in the industry, it's still largely tooled toward old-growth. So that's why there is time for a phase out. Any government decision is going to have a timeline, and we're saying the government has to commit to that timeline. But they won't even commit to that right now. They're just going to have the industry transition to second-growth when they're done the old-growth. We're saying, "Let's do it sooner, while we still have some standing big trees."

One of the quotes I read from you was about how Facebook changed environmentalism. Can you talk about that?
Facebook is like email on steroids. It's a very powerful tool because people are largely visual creatures. And we're also social creatures. So Facebook adds in the elements of photos, along with being able to send messages. People being social animals and also feeling disempowered many times in these movements— not everyone's an optimist like me, or confident we're going to succeed in these campaigns—are less prone to support things if they think they're going to put in the effort and still the government's not going to listen.

But when they see it snowballing and at first 10 of their friends are coming, then 100 of their friends, and then thousands of people are coming to these things—and they can watch it climb on a Facebook Events page, in terms of who's said, "Yes, I'm attending"—then it's more empowering and they feel like there's a chance to turn things around. So Facebook has certainly played a role in this recent upsurge as well, but it's not the only thing going on.

What are some of the other factors?
The other factors, and this is the hard thing, is the slow, difficult, painful repeated efforts to raise environmental consciousness. Essentially, it's constantly giving slide shows. I've given slide shows about four or five times every month for 10 years now about old-growth forest, throughout Victoria and throughout Vancouver Island. And taking people into the woods. So I've been doing hikes constantly to these different ancient forests.

Over time people see the news articles, they come out and see us, they come out and see the old-growth forest. And I believe movements grow exponentially.

So for the longest time you put in a tremendous amount of

effort and the number of people who are supporting just slowly and incrementally inches upwards.

△ Do you think there is something that has activated people's consciousness?

Al Gore. I don't think Al Gore created this current upswing in environmental consciousness, but he certainly was one of the catalysts to trigger what was coming anyway. Grassroots movements for climate change and environmental activism have spread all across the industrialized world, and when we're out of this recession in another year and a half, two years, you'll see this movement explode right through the ceiling. So revolution is just around the corner.

△ When I was thinking about what was going on in the Clayoquot Sound and various other forestry campaigns that I grew up around, I always thought about saving the old-growth forest as a good thing because it *was* a good thing—biodiversity and cute animals and pretty trees—and now people are making the connection that if you don't save the forests, we stop breathing. That's a very different way of looking at things.

There has been research to back that up as well. One of the things you'll hear some of the dinosaurs in the logging industry say is, "The best thing we can do is cut down the old-growth and replant healthy, new, vibrant trees that will save the atmosphere." It turns out that's wrong, actually. The fact is that old-growth forests store more carbon—which is taken from the atmosphere—per hectare than any other ecosystem on the planet.

When you convert old-growth forest to second-growth tree plantations, the ensuing tree plantations don't resequester the amount of carbon as the old-growth forest because second-growth forest only stores one-half or one-third or the amount of carbon per hectare as the original old-growth.

And you've lost the original carbon as a result of the decomposition of the toilet paper going out there, the pulp and paper waste, lumber products in the garbage dumps. There is a small amount, but it's a negligible amount ultimately, stored in longer term wood products like some of the lumber in furniture and so forth, but that's only about 8 percent of the carbon in the original old-growth. The rest of it ends up in the atmosphere, and a smaller portion is resequestered by second-growth. But second-growth just does not have as much carbon stored as the old-growth. That's a simple fact.

▲ **How do you change the perception of government and society to realize, "You know what, that wood actually has more value to us as carbon storage than it does as a table?"**
There are different roles for people in conservation initiatives; there are different ways to reach people. The scientists play an important role by bringing forward the hard facts in the research. And at the same time people need an emotional connection to these places too. There are no more determined advocates for old-growth forests than those who have actually seen Carmanah, hiked the West Coast Trail, been to Cathedral Grove, Clayoquot Sound.

▲ **But also I don't know how easy it is to save Cathedral Grove if it's called Timber Lot 102. I remember when the Great Bear Rainforest got named, and I thought, "Oh, that's genius. I can see it being a lot easier to organize to save the Great Bear Rainforest than to save That Really Big Patch of Land."**
The Central Coast Timber Supply Area. The governments complain about this sort of thing, but they have a fabricated name as well. A tree farm licence is also completely fabricated. The point is you call things according to how you believe they should best be depicted. But it's never the thing itself. And governments and

corporations fabricate their own technical names for liquidating those ancient forests.

▲ What can and should people do?

Often people say, "There's no magic formula to saving the environment." You know what, there actually is a magic formula. It's a very basic one. It's the maximum education and involvement of the maximum number of people. We believe that it's an educated citizenry that exerts the ultimate lobby pressure on government. That's why we spend the majority of our time on constantly informing people, and all types of people. It can't just be your typical university tree hugger. It's going to have to include people in faith groups, business owners and a wide diversity of different kinds of people from different ethnic groups.

Classic environmental activists by themselves are just a small fragment of society, which is why conservatives who believe in conservation, business owners, folks who go to church regularly, a wider diversity of people needs to be brought into the fold of conservation because everyone ultimately has a stake in this planet.

So the best thing people can do, first of all, is to get informed on the issue. Secondly, tell the government what you expect. And thirdly, spread the word to others, and one way to do that is by volunteering with organizations like the Wilderness Committee.

▲ When was the first time you saw an old-growth forest?

The first time was in pictures. A turning point in my life was when my dad bought me the *Natural History of Canada* series, which is like a picturebook from the 1960s of different ecosystems of the country. I saw this picture of six couples dancing on a giant cedar stump, maybe 150 years ago, with a guy playing the fiddle. And I was like, "That's incredible. All these people can dance on

a dance floor made from a giant tree stun

living in Saskatchewan and I couldn't im

big. In Saskatchewan you hug a tree witl

when I was 10, I asked my parents to t?

forests in British Columbia. We stop'

trail in Mount Revelstoke National Pai ҡ ⌐

Highway and that's where I saw my first big ceɑɑ ᵤ.

went to Pacific Rim National Park and Cathedral Grove. I was

totally a determined advocate from then on.

⌃ **If somebody walked over from the legislature tomorrow and said, "Ken, we want to put you in charge of forest policy for BC," what would you do?**

[Laughs.] I would transform things greatly, actually. First, I would enact concrete timelines to put an end to old-growth logging where old-growth is scarce, which is in the southern part of the province. Then I would ensure that the system of logging for second-growth is done in a more careful, sustainable way—more selection logging, no logging on steep slopes, large buffers for salmon streams, lower rate of cut. Then I would ensure that we ramp up, in a big way, the value-added sector side, partly by banning raw log exports as a guaranteed log supply. At the same time we'd have to restructure the industry so it's not based on these tenures for giant corporations, so you get more community forestry and woodlot licences and that sort of thing. So there would be a lot of big changes.

Generally what we have to do is slow down the amount of cutting in this province. It has been over the top. The recession has knocked down logging for a while, but it will be back up again by the time there's an economic recovery. Sustainability ultimately requires that we do more with less. We just can't keep taking as much as we have.

...o you feel about trees?

...ghs.] Obviously I'm a big fan of trees. They're a spectacular ...eature—a living organism that can live to be 2,000 years old. In fact, in Norway they found the spruce trees there can be 6,000 years old, which is insane, a 6,000-year-old living creature! And they're huge. It's hard to believe that a living creature—one of them—can be as big at those sequoias down in California or even the cedars over here.

I just love living creatures, and I love the diversity of living creatures. I always gravitated toward monsters when I was young because monsters are a breed of biodiversity. And some of these creatures are such fantastic shapes and forms, they definitely get my imagination going.

Western Canada Wilderness Committee at www.wcwcvictoria.org

5
JACKSON DAVIES
BEACHCOMBER

For almost 20 years, *The Beachcombers* defined the way the world saw British Columbia. The show is so iconic that even if you never saw CBC-TV's long-running family drama, you probably still remember it.

In 1990, I was lucky enough to interview actor Robert Clothier (a.k.a. the show's ever cranky not-quite villain, Relic) about the final season of *The Beachcombers*—which wasn't just Canada's longest running series, but was challenging *Bonanza* for the title of longest running dramatic series ever. And even though I'd never been a fan of *The Beachcombers*, Clothier's passion for the show was so genuine I found myself missing it desperately and furious at the Torontonians in CBC's head office who'd taken a chainsaw to part of Canadian culture.

The series also starred Bruno Gerussi as the hairy-chested beachcomber Nick Adonidas, one of the most iconic characters in Canadian television history, who was always working alongside his native partner, Jesse Jim (Pat John).

Jackson Davies (who played RCMP Constable John Constable for 16 of the show's 19 seasons) has fought to revive the series ever since it left the airwaves. He has helped produce two TV movies that reunited surviving cast members and introduced a new generation of viewers to one of Canada's most mythic meeting places, Molly's Reach—and the not-so-mythical land of Gibsons, BC.

The show aired in 60 different countries, so for a lot of people around the world the image they have of Canada's forests, loggers and trees comes from watching Nick, Relic, Jesse and Constable Constable fight their weekly battles over drifting logs.

I met Davies in December 2007 at Listel O'Douls in downtown Vancouver to talk about the death of *The Beachcombers*, the death of real-life beachcombing and how the whole world came to Molly's Reach.

▲ **Can you explain beachcombing?**

Log salvage. They'd go out after a storm or something, or after a boom—which is the tugboats that pull all these huge big booms of logs to the sawmills and if some of them broke away they'd float out there and they were fair game. You could pick them up and sell them back either to the boom operator or even to the sawmills. You could do anything you wanted with them. It was that law of the sea: if it's abandoned, it's yours. So basically that was the premise of the show.

This guy [Nick] made his living as a beachcomber with his First Nations partner, and they would go out and try to find logs. They had the slow boat, *Persephone*. And of course everyone loved Relic, who had the fast jet boat.

The neat thing is—or it's probably not neat—is that all that's changed. These guys can't even do this anymore. Conglomerates have got together and taken over beachcombing.

▲ **When I was in the Cariboo and saw all these German tourists, I remember hearing that they were there because they'd fallen in love with BC watching *The Beachcombers*.**

The show ran longer in Germany than it did Canada. And whether they're urban myths or not, there were stories that realtors would get long-distance calls from Germans saying they wanted to buy a piece of property as long as it overlooked Molly's Reach and that little inlet there.

Any particular memories of working with trees, working around forests?

Oh yeah. The neat thing was there were a lot of episodes that had to deal with logging. There was one really interesting episode directed by Charles Wilkinson—a very good local director—about a family that was in trouble. They didn't have any money. And the father basically started fires. The old proverbial idea that people would start fires in order to get a job fighting fires—it happens a lot. I didn't realize that, but it does happen a lot.

It was an interesting episode showing not only the obvious—the mistake about it and how the character was wrecking an industry—but also, very subtle for a *Beachcombers* show, that this guy wasn't just some guy smoking dope in the woods. It was a guy who purposely set a fire so he could get a job.

We got into a period there where we would do a lot of shows about the logging industry because it was big up there. That was how the *Persephone* would make money—getting those logs that went away from the booms. At that time, the beachcombing laws were different. Five bucks would get you a licence—that's all it cost you, five bucks and you could go out and make a living on the water basically dealing with timber that broke away from the boom!

Any other thoughts about how *The Beachcombers* spread the image of BC around the world, how it defined us?

I think the other parts of the world loved it for the scenery. They couldn't believe that this gorgeous place was there. American TV shows were either westerns, or someone being shot, or a sitcom, which was inside. So it was very rare that you got to get outside and see the beauty of an area.

How do you feel about trees?

I'm a big fan of trees. I grew up in Alberta—we didn't have a lot of trees. I had great relationships with trees. I was back home, and we walked by the park and I had this tree that I used to hang out in. It was an evergreen and some parts of it had died, so I could actually climb up in it and it was like a little turret, like a little castle that I could look out of.

As a kid, that was my favourite tree. I'd go and sit up there and look out, like I had a picture window out of a tree. At the time it seemed huge. I'm sure the thing was like 10 feet [3 metres] tall.

We had a place on the North Shore in West Van that was up high in the hills, so we were above the treetops. And when it would snow, that would be very cool, because you'd feel like you were on top of the world.

I haven't dated one, but I certainly appreciate them.

▲ **I know what you'd do if you ran the CBC or the Department of Canadian Heritage [which handles culture], but do you have any profound thoughts on what you would do if somebody said, "Hey, run the forests"?**
I don't know a lot about it. Obviously, it has to be sustainable. Actually, I was thinking about that the other day—thinking about bamboo and how everyone is putting bamboo in floors because the turnaround is a lot faster with bamboo. And I was thinking, "I wonder if we could grow a lot of bamboo?" I had bamboo at one place we had and it took over everything.

But obviously you've got to look upon it as renewable. If you don't, well, it's a huge industry and you've completely destroyed it. I have the same feeling about the lumber industry as I do about the Canadian film industry. I think you should add value to it. I'm a little upset with us just being workers for the big multinationals that come and shoot here. I would far rather own our industry and export our talent, and it's probably the same way with the lumber industry, probably just being able to add value

to the lumber industry makes sense. You ﹐
off. If you're going to do anything, use the
some great artists who do some wonderfu

Bruno loved it. You know, Bruno love
he passed away, someone from the First N
huge, big, wonderful piece of cedar. He a
house—because he had this incredible little   ﹍﹍at
he used to use—and he moved that out, put it in his wonderful
little shed right beside his house out at Gibsons and was going to
do some carvings in it. He used to love carving.

The Beachcombers at **www.thebeachcombers.ca**

6
VELCROW RIPPER
SACRED ACTIVIST

Velcrow Ripper started making films to change the world in 1979 when, at the age of 14, he wrote, directed and produced his first documentary, *Iran the Crisis*.

Since then he's made more than two dozen full-length documentaries, including the Genie Award–winner for Best Feature Documentary *Bones of the Forest*, where Ripper visited BC's logging blockades and spoke with Aboriginal and non-Aboriginal elders about the future of the forests.

His 2007 film *Fierce Light: When Spirit Meets Action* explores what Ripper calls "sacred activism" and took him on a five-year quest to find hope in the scared and the sacred places around the world, including the Killing Fields in Cambodia; Bhopal, India; and Auschwitz. "Places where some of the darkest days of human history took place and in those places I looked for sacred stories, stories of transformation in the face of darkness."

Fierce Light isn't officially a sequel to his hit documentary *ScaredSacred*, but it comes from the same sense of curiosity, passion and optimism that led Ripper to look for—and discover—hope, life and love in some of the most tragic spots on the planet.

When we spoke in September 2007, *Fierce Light* was about to start playing festivals and he'd just started work on a new film, *Evolve or Dissolve*, "about the idea of the evolution of consciousness and the evolution of the planet."

I met Ripper in his study at his home on Toronto Island—a room with an altar made up of spiritual icons he's collected from around the world. Asked to define his own spirituality, he replies, "I used to call myself a Sufi-Buddhist-Bahai-punk rocker. Currently I am convinced that any belief system would be a limitation."

We talked about Ripper's experiences with native and non-native elders, the fight to save a community farm in LA, nature-deficit disorder and the philosophy he calls sacred activism.

Could you explain sacred activism?

I discovered, as I started looking into the idea of sacred activism, that it was actually a *Zeitgeist* going on in the world, that old models of activism were beginning to wear out for people a little and that here was a new approach emerging—a new approach to activism that wasn't just founded on anger, that wasn't just based on reaction, but was actually based on response. A kind of activism that was about focusing on what you're for, more than what you're against. A kind of activism that recognizes the interconnectedness of all issues. It doesn't say, "My issue is the best issue," like "I'm into the anti-nuclear movement, so why should you bother being a feminist? The whole world is going to blow up, so feminism doesn't matter. Just do my issue."

Or an environmental movement that is rife with racism and sexism, but it's okay, we're going to save the trees.

There's this tunnel vision that activists can sometimes fall into, which I think is old paradigm activism. I think new paradigm activism, what I'm choosing to call sacred activism, recognizes that it's a holistic process. Just like the planet is interlocked, interrelated systems, nothing is separate. It's the same thing with activism.

BC has been the birthplace of so many activist movements. Are any of these sacred activist movements coming out of BC?

I think any activism can be sacred activism. Anything we do can be sacred or compassionate activism. It's just how we go about it. So, definitely.

The environmental movement, which is so strong and I spent many years making films about it, definitely has a lot of those components of sacred activism in it. I wouldn't make it black and white, like, "This is sacred activism and this isn't." I think it's part of a slow evolution. But I know in the time that I was in the environmental movement, it was still not recognizing the full interconnectedness of issues. But if any movement is suited to becoming sacred activism, in fact it is the environmental movement because it's an ecological model, the model of interconnectedness.

You've got Gaia as a model to work from.
Yeah, Gaia, exactly. Look at the forests and how the microrisal fungi keep the trees alive. A tree is not a single tree. A tree is interconnected with all other trees with this microrisal fungi. It's all interwoven. It's the same thing with the planet as a whole. There is no somebody else's backyard.

Global warming is an incredible unifier because it cuts across all boundaries, all nations, and it's like, "Oh, we're all screwed." That has unifying potential. So that's one aspect of sacred activism and there are many.

Another one is that it comes from the heart, not from the head so much. There was an old model on the left: "If we just pummel them with facts, if we just get enough facts, that's going to save the world." But you know what? That doesn't save the world. We're inundated with facts. We're inundated with information.

And traditionally if you look at elections, the people who come at you with facts tend to lose the elections.
Now if you're looking at the world today and talking politics, someone like Barack Obama is coming from a position of sacred

activism to some extent. He's echoing Martin Luther King in some of his approaches. I think he is actually a representation of this *Zeitgeist*. People are hungry for this kind of an approach. People are tired of cynicism on all sides. They're tired of polarization. They're tired of us-versus-them politics, us-versus-them activism. This idea of interconnectedness is so key because when you take it to the furthest extent—like the Buddhists do—there is no other, there is no separation.

In *Fierce Light*, we go to South Africa and interview Archbishop Desmond Tutu, who talks about the idea of *Ubuntu*—and *Ubuntu* is the idea that I am because you are. A single isolated human being is actually an impossibility in the same way that a single tree is an impossibility.

▲ **Where else did you go for *Fierce Light*?**
One of the main central stories is about South Central Farm, which is in Los Angeles. In the wake of the Rodney King riots 15 acres [6 hectares] of land was given to the Latino community, and low-income families were offered a plot of land to grow their own food. And in that 15 years since the Rodney King riots, this incredible Garden of Eden grew up right in South Central LA, right in gang-land LA. And it was a largely indigenous population, so they're growing heritage foods. Food anthropologists went down and checked this place out. It was just really a miracle. The largest urban garden in North America.

The city sold it to a developer in a backroom deal, it was a previous administrator. And they didn't tell the farmers. When the farmers found out, they tried to fight it in court, but it was too late, they couldn't get it back. And the developer refused to sell it to them. They put a call out, had a big occupation to save the farm.

Daryl Hannah was tree sitting. Julia Butterfly was tree sitting. And I spent a month down there. It was actually Daryl who called me and got me there. She's a *ScaredSacred* fan. It was a remarkable occupation. It had the best food of any occupation I'd ever been on. It was all this delicious organic, heritage food. People from all over the world came. They put a call out to spiritual leaders. There were different spiritual leaders every day. There was a prayer vigil every night, circling the farm. We were living in this beautiful Garden of Eden. And it was just so inspiring. The indigenous culture really flowered during the occupation—there were Aztec dancers and ceremonies. To try to save the farm, one traditional ceremony—one they hadn't done for years and years and years—went on for 18 hours.

In the end, they did tear the farm down. The developer refused to budge, even though they raised $16 million.

But what's amazing is that the farmers have not given up. They've actually rented a community centre kitty-corner to the farm and every month they have a farmer's market. They bring in organic food to the neighbourhood. They have traditional Aztec dancers. All these bands have formed, like Farm Life that does hip-hop, from the occupation.

And they're still going strong, and they're not going anywhere. And no one wants to buy the land, it's sitting vacant. [The developer] bulldozed the gardens, the gardens are gone, but [the community is] determined to get that back and replant.

▲ **Can you talk about some of your environmental films like *Bones of the Forest*?**

Bones of the Forest is a documentary finished in 1995. It's a 16mm feature-length film. It took five years to make. We lived in the old-growth forest. We lived in the Walbran Valley on the blockades and we lived in a houseboat in Clayoquot Sound and on the blockades for years.

It is a story of elders. You had to be 70 or older to get into that film. It was all native and non-native elders talking about the forests and the struggle to save the forests. It was such an honour just to meet all these incredible elders, these great repositories of wisdom who are often so neglected in our society. I think that was one of the greatest gifts of *Bones of the Forest:* this opportunity to meet these amazing people who are so marginalized so often. It was also an honour to be able to go into the native communities and to be welcomed and to meet them. We were really, in that film, trying to heal a split that was quite real in the environmental movement between native and non-native activists. There is a tendency in the environmental movement to see native people as potential allies, but also potentially in the way to saving a park, or creating a park or something.

There hasn't been—I think it's changing—but there hasn't been this sense of real, deep understanding of the issues that native peoples care about. And it's not so simple. And you really need to go with an open heart and an ability to listen, as opposed to fixed ideas.

▲ **Because nobody is going to be able to find this film since it isn't available on DVD, can you tell us what people who watch this come away with?**
I think you're going to come away with a sense of the beauty of the old-growth forest. You'll come away with a sense of the spin that the logging companies put out. One of the characters is the vice-president of MacMillan Bloedel, and you just see how they start to believe their own rhetoric.

There is this line in the film that is quoting Goebbels, the PR guy for Hitler. He says, "Never tell the truth if a lie will do. Always tell a half-truth if you can, because a half-truth is better than a lie; you can't prove it, or disprove it." And "If you tell a lie often enough, even the liar will begin to believe it." Those are the principles of greenwashing.

Merve Wilkinson says it, talking about forestry PR.

▲ **Isn't Merve sort of an environmentally friendly logger?**
Yes, he's criticizing greenwashing.

▲ **So he was explaining the concepts of greenwashing?**
Yeah, which is when logging companies or environmentally destructive companies hire someone like Burson Marsteller to come in and make them look good. [Burson Marsteller is a multinational public relations company known for managing corporate crises like the Bhopal disaster and the *Exxon Valdez* oil spill.]

▲ **My understanding is that Merve was one of the first environmentally conscious loggers.**
Merve is in the film as a sustainable logger.

And that's another thing you'll take away, is that it's possible to actually log and enhance the forest. If you look at Merve's Wildwood land, where he's logged for 40 years, it has all the properties of an old-growth forest. He selectively logs it, and he gets a yield from that that's sustainable for him. So you can log in a way.

Clear-cutting is just greed, short-term greed, because these companies have no allegiance to the local economy. The global mindset is that we'll just use up this forest, then we can just go to Brazil. We'll finish off BC and then we'll go to Brazil. There is no commitment to sustainability.

▲ **What were some of the other things you came away with covering the front lines of the environmental movements in BC?**
I got a real sense of the dedication of these activists, people willing to put themselves on the front line. I loved the elder activists like Ruth Masters, one of my heroes, from Courtenay. She must be in her late 80s now, but Ruth is somebody who—she knows what's right, she knows what's wrong. And she's talking

to those cops like they're little boys: "Now listen here, s[...] you cared about the environment, you would not be arres[...] me. Would you?"

And she plays her harmonica and plays "God keep our land glorious and free" from "O Canada" as she's being arrested. She sort of squeaks it out.

I came away with the sense of the pricelessness of old-growth forest. There's 3 percent left—I don't know what it is, 2 percent left? They are 10,000-year-old works of art that can never be replaced. This is something we really need to get into our heads. A tree farm, even a restoration attempt—a multispecies restoration attempt, which is very rarely done, it's mostly just tree farms they replace clear-cuts with—has nothing like the incredible grandeur of these old-growth forests. You can just feel the primeval quality of them. And the film captures that, to some extent, because of its cinematography. Then the sense that the native communities are the best stewards for this land.

Why do you feel that way?

Just intrinsically. Now it's complicated because we're dealing with a colonized people, but I feel that way because they traditionally have been, because they're the ones who often live there. Given the right circumstances, they'd be able to use the land sustainably because traditionally they have known how to use the land sustainably. Time and time again, we're seeing that when there is a healthy collaboration between the environmentalists and the native community, there is great power there, enormous power.

How do you feel about trees?

Oh, I love trees. Trees are remarkable. Trees are—they move slower than us, they have a different time frame. Especially 1,000-year-old trees. They're just amazing.

...one of the characters in the new film, and ...ge of her after two years of sitting in that ...er coming down and her feet finally touch-

...t wonderful metaphor for the environment ...self is so critical to our well-being. We're suffering from a nature-deficit disorder in our society today. It's a serious issue. Kids are growing up having never climbed a tree. And it's really affecting our psychological health.

If somebody put you in charge of forest policy for Canada, what would you do?

I would eliminate unsustainable practices. And I would say we have to consider seven generations—and seriously consider it. Is there going to be a forest here seven generations from now with these practices? If not, we need to change our practices.

Velcrow Ripper at www.fiercelight.org

7

JAY DODGE
TREE SPRITE

I'm scared to death of heights. So how did I find myself on a tree platform 50 feet (15 metres) in the air? The answer . . . Jay Dodge. Dodge is one of the founders of Boca del Lupo, a Vancouver-based theatre company that became internationally famous for doing site-specific work in Stanley Park, starting with *The Baron in the Trees* in 2002. *The Green Chain* features a tree sitter, and Dodge and his partner, Sherry Yoon, were the only people I was prepared to trust with putting my actor—never mind me—up a tree.

Dodge is an actor, writer and director, but once you see him in his element—skittering up trees and bouncing among branches like Tarzan—it's clear he's actually a tree sprite.

He first learned to climb working on sailboats: "Often you find yourself up at the top of the mast or dealing with all the ropes and rigging and the same kind of thing that mountain climbers use." During the five years he spent treeplanting in the interior and northern BC, he connected with mountain climbers and that eventually led to climbing trees.

That tree I climbed was in his backyard. And it was located next to a ravine so that even though we were only 50 feet (15 metres) in the air, when we looked down (or filmed down) it was closer to 100 feet (30 metres). Great for the movie. Extra nerve-racking for a guy who wouldn't go up the Eiffel Tower or stand too close to the edge of balconies. But I knew that Dodge wasn't going to let me fall.

I talked to Dodge in May 2009 about his life in the treetops and about visiting Vancouver's iconic Stanley Park after the 2006 storm that took down so many of the park's oldest trees.

▲ **What was it like going into Stanley Park after the storm? I'm assuming you'd bonded pretty seriously with those trees?**
It's funny, I heard about the storm on the news, but I didn't think much of it. I'd treeplanted for years, I knew blowdowns happen. I've been caught out in a storm on a cut block where huge trees are falling and you can literally feel it like an earthquake, the ground rumbling underneath you. It's quite an awesome and terrifying experience. But I also know that it's quite natural. It happens.

So I didn't think much about it until one of the actors called up and said, "How are you guys?" They were worried for us, like were *we* going to be okay because our trees might have been blown down. And that's when it hit me: "Oh yeah, our favourite trail was right where most of the blowdown had happened." Or very close to it. There were trees, some of the trees we had bonded with, some of the trees we had names for, and it was quite something when I finally went down there and drove the car around these areas where it used to be lush, beautiful, thick forested areas—they were gone. You could see straight out to the ocean. When I actually climbed over all the blowdown and got into the area of the park that we had used most often, all of our favourite friends were still there. We were just on a little ridge that helped protect them from being taken out so most of the trees that we had come to know and love were not part of the blowdown.

But it affected us in that we still haven't done a show in the park since that happened. The park staff have taken a long time to figure out what they're doing and decide how they're going to rearrange things in there and clean things up. So we're hopeful that one of these days we'll get back into the park and do another show there.

▲ **What's the highest tree you've ever climbed?**

That's a good question. I don't know. Often when we're rigging up some of the things we do in the trees, you have to go a lot higher in order to come down onto something and make it happen for the production. I guess we've probably gone up 200 feet [60 metres]. I don't think it would be much higher than that. That's a pretty high tree.

For someone like Julia Butterfly Hill, some of those trees, I don't know what the heights would be, I'm sure they get up to 300 or 400 feet [90 or 120 metres]. With us, it's probably 200 feet, and that seems plenty high as far as I'm concerned.

▲ **That seems plenty high to me.**

It is quite a feeling. You're up there and you feel the tree move. When you're down at the roots, the tree feels pretty sturdy, and it's sturdy still when you get up top, but it starts to sway and move and flex and it feels almost more like you're on water when you're up that high. It's quite a wild feeling.

▲ **If somebody put you in charge of our forests tomorrow, what would you do?**

I would certainly want to save any old-growth that's still out there. I think it's irreplaceable. Near my house here on the Sunshine Coast there's a tree—one old-growth tree—that has survived being logging and survived the last forest fire here about 100 years ago. It's one of those trees that takes eight people to get around it and it's phenomenal. It's one tree that people come and visit, this one tree that survived. It's awesome and it's becoming more and more rare.

I cherish trees, it's part of the reason I love being in British Columbia. But I also understand that trees are part of a lot of people's livelihood. I think there are ways to have sustainable logging and treeplanting. In my last years of treeplanting, for instance, when I was a better planter and logging practices had changed, there were all sorts of different techniques that really did seem like they were sustainable ways to deal with things. They were going back to horse logging, where there was low impact and just grabbing certain trees. I understand that it's a complex situation. I don't think it's just about never cutting down another tree. And I don't think it's about just letting logging companies go ahead without any oversight. It's somewhere in between. We're tied to it.

I live between two worlds—I live rurally and I also work in the urban centre of Vancouver—and so I get to see both sorts of mentalities. You realize there is a huge interdependence among the rural and the urban communities of British Columbia. They have things to learn from one another in terms of finding a way to both be sustainable, but understand it is a resource.

I guess if I was put in charge of it I would want to—I'm horribly underqualified to even answer this question—but I would want to set an example for the world of what it means to have a sustainable forest industry, as well as conserve the most beautiful and oldest and most vibrant and lush parts of our beautiful province.

▲ How do you feel about trees?

I love trees. What's not to love? I'm looking out in my backyard and there's a grove of cedar trees that the last owner didn't cut down, thankfully, and it's beautiful. I tend the grove. It's one of the most cherished spots on my little plot of land here. Trees are fantastic. They're wondrous things.

Jay Dodge and Boca del Lupo at www.bocadellupo.com

8
SEVERN CULLIS-SUZUKI
SECOND-GENERATION ACTIVIST

Severn Cullis-Suzuki is from Canada's first family of environmentalism. At the age of 12, Cullis-Suzuki rocked the 1992 United Nations Earth Summit in Rio de Janeiro, stealing the spotlight from global leaders like Al Gore with a speech that still circulates the planet on YouTube. "In my life," she said, "I've dreamt of seeing the great herds of wild animals, jungles and rainforests full of birds and butterflies, but now I wonder if they will even exist for my children to see."

Since then she's been doing her part to keep those jungles and butterflies alive. She's on the UN charter commission, she's in *Vanity Fair*'s eco hall of fame, she's got an undergrad from Yale in ecology and evolutionary biology, she has a master's degree in ethnobotany from the University of Victoria, she's one of the writers and editors of the inspiring book *Notes from Canada's Young Activists: A Generation Stands Up for Change* and, yeah, her dad is Dr. David Suzuki.

But Cullis-Suzuki is not just recycling her dad's ideas, as she tries to redefine activism and environmentalism for a new generation. I talked to Cullis-Suzuki in July 2007, just after the release of her book, about taking responsibility, ancient agriculture, pulp mills in Tasmania and why she doesn't call herself an environmentalist.

In the intro to your book, you talk about your passion for the connections between culture and the ecology. Can you talk a bit about the connections you see?

That's a big question. I think we are very connected to our environment, and it's not just in a physical way. I mean, it's obvious when you think in biological terms how connected we are with our environment just from the simple fact that we eat food. Food is where the external world, your environment, becomes part of your body, and this is something we don't often think about. But besides that, we have all kinds of values and experiences that really shape who we are that simply come from the matrix where we all exist, which is another way of describing our environment. I think most people don't really think of that when they think of this word, *environment*; it tends to be marginalized or compartmentalized as having to do with crazy old David Suzuki, or maybe the Kyoto Protocol, or maybe hippies chaining themselves to trees or something that is very much externalized when it's simply all that surrounds us. And we're very culturally affected by the culture of the environment that we surround ourselves with.

▲ **I read some interviews where you don't identify yourself as an environmentalist. Can you talk about why?**
I just feel like I'm really a concerned citizen. I find that when I hear the word *environmentalist*, I tend to have a certain stereotype in my mind. You think of the word *environment* and your mind immediately goes to recycling, or some image of a natural park, or something that is out there, that isn't relevant to your daily life.

And I'm very concerned with human welfare. I'm very concerned with my own well-being. This is why I care about these so-called environmental issues because they really are social issues. They really are about quality of life. They're really about all these other issues that, to me, is what it's really all about and not necessarily about panda bears and whales. That's not a very good way of describing what I mean, but we just need a more holistic way of looking at the world around us.

⚠ Do you have a description you use? Am I supposed to call you an activist?

At first we were a little scared about calling our book *Notes from Canada's Young Activists* because a lot of people in this book wouldn't necessarily call themselves activists. I think that word, like *environmentalist*, can be whatever you define it as. And I really feel like an *engaged citizen*. Or *advocate* is something that I'm more comfortable with because I feel like I want to encourage people to speak out and be active in their community, but *activist* has a lot of baggage, and I really believe in being positive.

⚠ Can you explain what ethnobotany is, beyond one of the coolest words of all time?

Well, ethnobotany is in a field called ethnoecology. It sounds complicated. It's really simple—"ethno" is humans and culture, and "botany" is plants. I've been working with Kwakwaka'wakw elders learning about their plant-resource management and how people used to gather plants.

It's a relatively new field. It started in the early decades of the last century. It's one of those small disciplines that is growing. And for me it's a field that is absolutely essential as we start to talk more about sustainable resource management. It strikes me as crazy—if we're trying to figure our how to live and sustain and produce in the 21st century in a sustainable way—how we could not at least have an understanding, or ask people who lived in the same place for thousands of years how they did it, how they lived sustainably, how they sustained tens of thousands of people in different areas of the province. We've got to know how they did it.

▲ **I thought it was fascinating that in your essay you dealt with historic aquaculture.**

It's an amazing story, and when I found out about it almost three years ago, I was absolutely blown away and I think that was a real turning point for me. I was sitting in [ethnobotanist] Nancy Turner's class [at the University of Victoria], and she'd invited two [First Nations] people to speak to us about clam gardens. They were describing these traditional aquaculture clam gardens. There are these terraces that people used to build up and down the coast and you can still see them, they're very distinct, they're around flat sandy beaches and where you find a lot of clams. These were manmade—they were made by people for who knows how many thousands of years. They rolled rocks down to the zero-tide level—which aquaculturalists are discovering is the optimal level for clams—and they were actually creating more habitat so that people could eat more clams. And if you go back in the middens, apparently about 2,000 years ago the shells shift from being mostly oysters and mussels to being mostly clams. They think that's when the technology for clam gardens arose so that it could be a major food source to sustain the many people that lived in that area between Vancouver Island and the mainland.

▲ **Did you discover anything about forestry that's applicable now?**

One of the main concepts I've been learning about is, as Nancy Turner calls it, this attitude or this ethos of "keeping it living." She calls it "keeping it living." It comes from her learning from the many different elders she was talking to, specifically Kwakwaka'wakw teaching. And the concept is pretty basic when you really think of it, it's harvesting plants, harvesting resources in a way that keeps that resource alive, that keeps that population abundant.

You can see this ethic when you look at CMTs—culturally modified trees—and you see these giant trees that have been alive for sometimes 1,000 years, and they have a big plank cut out of them, but they're still alive. So that would have been harvested in the ethic of keeping it living because things were harvested in a way that didn't fundamentally damage the resource base.

To us—living in the 21st century, coming from Western training—this is an amazing concept and that's because all our fundamental principals of resource management, of forestry, are just "take as much as you can." We're basically taking too much from our capital, which is why what we used to think of as being a renewable resource, like trees, now it's becoming not such a renewable resource. So that is a pretty good ethic that we should learn from.

▲ **Reading some stories about you I learned about you working to protect forests in BC, in Southeast Asia and the Amazon. But it wasn't specific what your involvement was. Can you talk a bit about what you were doing?**

I got involved in some of those projects abroad through my family. My family has been very involved in issues in the Kayapo, which is in Brazil, and working all over the country with different indigenous groups to learn more about their issues. So there was a lot of family involvement and support and protests. When I was very young I went down to the Amazon and it was a pretty life-changing experience for me, just seeing the kinds of impact that deforestation due to burning in the Amazon is having—not only for the Kayapo people, but for the poverty-stricken Brazilians who are actually inflicting this. It's totally unsustainable for everybody. I also started getting involved in Sarawak very young and I had lots of influence—

▲ **You're famous for getting involved very young.**

Yeah, exactly. One of my first projects was fundraising for the Penan people of Sarawak to buy them a water filter because the forestry that was going on at the time in the 1990s was polluting the water that a Penan group of people had to drink. So me and my friends decided we wanted to fundraise and buy a water filter for them, which we were able to actually present to some traveling Penan people. So that kind of involvement is the level that I've been involved at.

You told me you saw some things in Tasmania about how they are handling their forests.
I was recently in Tasmania to speak at a conference on biodiversity. While I was in Launceston, which is one of the main cities, I learned about this issue which was the building of a pulp mill, and the town was just at the point of deciding whether they wanted a pulp mill. Right now the biggest industry in Tasmania is the export of woodchips, and there's a lot of old-growth logging that's going on to make woodchips that get shipped abroad and then made into paper. One of the reasons for building a pulp mill is to have value added to the resources. I mean, it's crazy if they're just shipping out the woodchips and then buying back the paper.

So that was one of the arguments, and it became this big debate in the community. The community was very divided. But at the same time there was an amazing amount of really healthy debate. There were civic forums, there was a lot of attention in the media to the pros, the cons. There was also a lot of attention to a democratic issue that was happening at the same time there. We have the one issue that is basically a decision: pulp mill or no pulp mill. And there was the democratic issue where the major forestry company in all of Australia was putting a lot of pressure on people to support the building of the pulp mill and not necessarily pushing through the plans to ensure that the pulp mill was ecologically the best of the best.

So there were all kinds of weird intimidation things happening, and all kinds of silencing of people's voices. It made for a very interesting dynamic. I couldn't really say whether they should or shouldn't have a pulp mill, but all I could do was talk about our experiences here in BC and ask them to do their homework and ask other cities that have pulp mills what have been the ramifications to their communities.

Looking at BC, what did you say? What did you warn them about?

A big thing is health and health issues. I'm not going to name any areas specifically, but the health consequences of having a pulp mill in your town are quite major. People have to look at cancer rates, they need to look at asthma rates. Air pollution is already a problem in Tasmania, so they've got to think about that. And really look at the stats for what health problems are, because they might not be correlated by scientists necessarily, but you might just happen to have a high cancer rate in a small town that has a very large pulp mill.

Another thing they have to realize is the pollution of shell-fish, the pollution of their marine environment around that pulp mill. I don't know anywhere that has a pulp mill where people are still gathering clams and harvesting the food right around where that pulp mill is. It's just accepted that you don't gather food anywhere near a pulp mill.

So that has an economic cost, and those are the kinds of things the community has to be aware of and think about when they're making decisions. And finally, we have to fundamentally think about the long term, and we're really not doing that right now. We're really not doing that. It's something I bring up a lot. A big part of my role as a young person speaking out is asking people to really think about their children when they make these decisions like whether to have a pulp mill. People love their children. And if that's true, then they've got to make decisions for the future of their kids. I think that totally puts a different spin on the kinds of questions people should ask when they're making those kinds of decisions.

▲ **Can you explain what Recognition of Responsibility is and how it came about?**
Recognition of Responsibility came about around the end of my university experience. I went to school in the States, and I'd made lots of American friends—to my surprise, they're not all like George Bush. And our graduation was right on the eve of the Johannesburg Earth Summit, which was the 10-year anniversary of the United Nations Earth Summit in Rio de Janeiro. This anniversary got a lot of attention. They were going to have another summit. George Bush decided not to go, and my American friends were pretty upset about this. They really were concerned about how the world would perceive America. Of course, this was post–September 11, and there was already a lot of polarization between America and the rest of the world on how they reacted to that. And now here George Bush wasn't going to go to the Johannesburg Earth Summit and participate in this very important work.

So they wanted to send a message to the people and to the media at the conference that not all Americans are like George Bush and that many Americans are willing to make a commitment and take responsibility for their actions. We came up with the Recognition of Responsibility pledge, which says that: Today I take responsibility for my lifestyle. I acknowledge that I live in one of the most consuming countries in the world and my per capita—me—my use of resources and the waste that I produce is much higher than the global average. And today I'm going to try to take responsibility for that. It lists a whole bunch of ways that we can start dealing with our own ecological footprint. And the idea was that if we had a roster of people who had signed up for this, people wouldn't feel so alone, they wouldn't feel like, "Well,

if I take my bike to work it's not going to make a difference."
They would know that there are 5,000 other people pledging
to make this effort, so the effect is actually magnified. That was
really the idea and how it came to be.

**If somebody put you in charge of the BC forests tomorrow,
what would you do?**
I'd need like a week to think about that one. One thing that
has really confused me is why environmentalists and loggers
weren't on the same team. To me, someone who wants to
have a livelihood harvesting wood has to be involved in their
sustainability. I've never really understood that. I have a lot of
ideas why, and I think it's also a big task for the environmental
movement, the conservation movement, to start shifting their
ideas of what is "natural" and how to use forests, because we
all need paper, we all need to use the forests, but we need a
vision that can actually maintain these forests, not only for
their beauty and biodiversity, but also for their productivity and
for the resources. I just see that goal serving both people who
want to have parks and people who want to have jobs. It seems
crazy to me that no loggers who are currently working in the
industry today, none of them are encouraging their sons to join
their profession. And that's because they know that it's on its
way out. So we've just got to really start thinking about how
those two paradigms can come together because I think they're
absolutely joined.

How do you feel about trees?
I think that you'd be hard-pressed to find someone who didn't
feel some kinship for trees. I'm always amazed by how sheltering
trees are. If you go from standing on pavement, or standing in a
clearing, to standing in a stand of trees, there is such a marked
difference. You're either really hot in the sun of a clearing and

then you step into shade and it's just so relieving, or you're really cold and you're out in the open and then you step into a grove of trees and you're suddenly sheltered. Every time I experience that, I think about how obvious it is that we're cutting down all the trees on the planet and of course we're going to have some localized climate change because we're totally shifting these trees, which are so important for regulating atmosphere. So I think that they're this amazing force and I love trees. But I also think that they're going to outlast us. I don't necessarily feel sorry for the trees, I feel more hope that people will have respect for them.

More on ethnobiology at ethnobiology.org

9

DR. RICHARD HEBDA

HOT SCIENTIST

Richard Hebda doesn't just want to understand climate change, he wants to explain it to you.

The curator of botany and earth history at the Royal British Columbia Museum in Victoria, a trained botanist and geologist, Hebda's special area of interest is "the origins and dynamic of eco-systems and the distribution of species over long periods of time."

Hebda had the chance to explore his passion—and tap into the *Zeitgeist*—when he helped create an exhibit that shows a series of possible futures for the province of BC and the planet Earth—and a lot of those futures are the stuff of scary science fiction movies.

I met with Hebda in October 2008 at his office at the museum, where we talked about the speed of change, how modern manage-ment practices transformed BC's forests into fire hazards and beetle buffets and how the climate crisis may have created "the greatest teaching moment of all time."

▲ **Can you talk about what climate change has meant for for-ests, how it has affected the beetle population, how it has affected forest fires?**

In the context of forestry, climate change is a very big motiva-tor or mover. The first thing that we already know is that there *are* consequences from what people would say are minor cli-matic changes in British Columbia. These are the mountain pine beetle epidemic on lodgepole pine, which has now moved onto ponderosa pine in the Kamloops area, and a needle blight called Dothistroma, which is a response to warmer but damper climatic conditions in northwestern British Columbia.

We know from our climate models, when we model the impacts of the changing climate on forest distribution—and we show this in our gallery—that the distribution of forest ecosystems will be less.

So not only will there be different forests, which will function differently, but there will be different distribution of forests. Particularly in the southern part of British Columbia, there will be less forest. In the north, there may be more because it will climb up into the higher elevations.

You're talking in scientist time. You're talking in centuries, right?
No, I'm actually talking about changes in forests over the next few decades—by the end of this century. For example, the ponderosa pine forest in the southern interior near Kamloops has pretty much died out—there are a few smaller trees, but that open forest parkland is now gone. I hope the climate is suitable to allow the smaller ponderosa pines that are growing there now to develop into big old trees. But our climate models suggest that this may not happen because it will be too warm and dry for the trees to grow to maturity. So changes like these are already underway.

How do we ever get rid of the pine beetles if we're never going to have another cold winter?
The beetles will eat until there is no more food for them to eat. Basically, it's the classic boom-and-bust situation. We don't get rid of them. They were always here. It's just that they got a leg up on the trees. We have a large population of uniform-age trees, relatively large trees. So once the beetle epidemic builds up enough intensity, enough individuals, the concentrations, essentially, of

the beetles in the atmosphere are big enough that wherever they go they find something to eat and they eat it until it's gone.

▲ **So we've created beetle buffets?**
We did. That's exactly it. Beetle buffets. And the other factor that people have mentioned that probably has a role to play is the summers became a little drier and warmer, so the forests are stressed. So there are beetle buffets in lodgepole pine territory. Elsewhere the forests are a little more stressed due to environmental conditions, so they have less capacity to deal with beetle infestations. There are so many beetles that there are clouds of them in the sky. Whenever they land on a tree of suitable size, you know—dinner. Eventually they eat everything up and they can't propagate themselves in large numbers, so they'll return to very low numbers.

▲ **What about other impacts of climate change? What about big fires, like the Okanagan fire [in 2003]?**
In terms of fire in the future, many of the models show—and the fossil record does too—that the warmer and drier it is, the more fires you can contemplate or have.

Fires are a natural phenomenon on the landscape and would occur one way or another. One issue that needs to be talked about is that we have suppressed fire and therefore we have fires that are prone to burn more intensely. So we have this question of how do we deal with issues of potentially more fires in a landscape that's been managed to increase the likelihood of more severe fires?

And then in parts of British Columbia they're using fire as a restoration tool. You remove the duff, you prevent the fire through the ingrowth, you prevent the fire from laddering, going up to the top of the trees. And you create an open understory. So that's one strategy in use.

But there can be arguments in some cases: If you do that to the entire landscape, then you've actually converted the landscape to yet some other state created by humans, so to which extent is that appropriate? And what's the normal distribution of ecosystems and burned and unburned areas on the landscapes?

I'm beginning to think—and a lot of people have considered this—that a mosaic is the appropriate way, a mosaic of stands of different compositions and structure. That would be the normal way in which the landscape existed, rather than one relatively managed, somewhat uniform set of circumstances for what people need.

Around urban areas there's a legitimate case to fireproof the area's buildings and around the interface through techniques that are consistent with good forest health, but don't overmanage the fire. So removing the undergrowth, the young growth, preventing laddering, perhaps removing a few trees so you have a crown a little more open and making sure you don't have tinder boxes next to your houses.

But out on the other parts of the landscape, there are a whole bunch of other values—the amount of forest cover and how that relates to water and how that relates to biodiversity. So it's a complex issue.

▲ **What is your answer for the future?**

For future forests, there has to be acknowledgement that—and I start all my talks with this—transformation is underway. They are not going to be the same. They just cannot be the same because the climate will change and it will change enough so that they will be different.

So what I've been asking people to think about—and this is very similar to what the Ministry of Forests has actually been thinking—is something called "forest resilience": how do you create forests that can change, but can change in their normal, natural way, into some future forests that are still healthy, functioning forests in those areas where forests have a good likelihood of persisting?

You need good biology, good ecology and good forest management, which means a whole bunch of things—how do you do silviculture, how do you remove the trees, how do you measure and set standards for biodiversity values and all these other kinds of values? How do you replant? What do you replant with? How do you even think about replanting future forests when you know they're going to be different? Can you bring in species from other places, and to what extent? How far away will those species go?

A suggestion has been made that, well, maybe we can plant larches in places where they don't grow, in very different parts of British Columbia, but where the climate is going to be the same as where they grow today.

And these are important things to think about from first principles. Everybody really has to understand the first principles of ecology. There are two things going on here in terms of the way we are thinking about our forests. One is that we know we have to do something, but we're going to tinker.

This isn't about tinkering. We're not going to tinker our forests into some sort of future because the mountain pine beetle, Dothistroma needle blight, or other things will take over and tinkering won't have mattered a whit. We will have wasted our time and money.

But the other thing is this: we really are talking about ecology in the broadest sense, where humans are part of the ecosystem and you've got to get the ecosystem principles right. What drives the forests? What are the key processes? What are the key species? How might they respond? What is our appropriate way to foster them or not interfere with them—in other words not increase the risk—in such a way that they can move into the future according to their own ways and means. Nature must have an opportunity to take its course while acknowledging at the same time that people have needs from the woods—in the woods and from the woods.

▲ **When you were showing me one of the climate change exhibits, it showed that one of the steps we could take to fight climate change is plant more trees. Except trees don't tend to be valued for the fact that they take CO_2 out of the atmosphere. How do we deal with that?**

Well, here's the classic recognition now, particularly in British Columbia, that forests are an important place to keep the carbon in the ground and in the trees. Photosynthesis takes carbon dioxide out of the atmosphere, which we seem to not think about enough. I think people in cities, for example, forget that there aren't many ways in which you can take carbon dioxide out of the atmosphere right now that work very effectively. Trees are the best. They've been doing it for a very, very long time, hundreds of millions of years. Algae in water—very important too and a solution to reducing the amount of carbon dioxide in the oceans. These are what we have in terms of our natural processes. And the one that we can influence and work with is the trees. You've got to keep them there. You can't just cut them down and hope that 50 years from now, when you replant them, they'll start taking lots of CO_2 out.

▲ **That's a major paradigm shift. I grew up in BC around protests to protect old-growth forests, and when you talked about protecting old-growth forest, the image was that you're protecting nature because nature's a good thing. There wasn't a sense that we're protecting nature because if we don't protect nature we're not going to have any air.**

Nature does a whole bunch of things for us, and let's not forget that we are the children of nature. We aren't here because we evolved in a rock somehow and emerged on the surface of Earth. This has been an ongoing, joint journey. It's only at this point where we have emerged out of nature, or we imagine we have emerged out of nature, hence we don't think that we could have changed the climate, but we have. We emerged out of nature and somehow changes in nature will have no consequences. In fact, that's not the case. As our exhibit says, "Climate rules." You can want whatever you want and imagine you can do whatever you can do, but you're not going to grow vegetables on Ellesmere Island, end of story. This is something people need to realize.

But in the context of forest ecosystems, so many gifts from the forests are vital to us—clean water, soil formation, CO_2 removal, timber, jobs, food, spiritual uplift, aesthetics—and it goes on and on and on. Not only clean water but a regular, reliable supply of water that then feeds the rivers, which produce the fish, etcetera, so there's no escaping it.

I think people have never understood this connection. And now an even more fundamental process is keeping the carbon. We've got to keep the carbon, man. We've got to keep it so that it's not going into the atmosphere, and forests do that. They not only take the carbon from the atmosphere and turn it into wood or decomposing material slowly in the soil or fungi or whatever you want to imagine, but they keep it there. They keep it there as part of the development of the soil of the standing biomass, the understory.

So this value, which now people are talking about in terms of dollars and cents because we're talking about carbon taxes and cap-and-trade systems, is now being recognized as something that has economic value. But that has to be done in the context of all the other values—maintaining biological diversity, water quality. All that means is healthy, resilient sustainable forests, which may change through time, but remain there doing their job as they have done for millions and millions of years. Other people think, "Let's just manage them for carbon." Well, it's more than carbon. Our future depends on all the values they provide, not just the little carbon bit, which is also very important.

▲ **It's hard to look at your climate change exhibit and not get at least a little terrified. How scared should we be?**
I'd rather put it as: how concerned should we be? When I speak to groups, they'll say, "Well, that's frightening." I say, "Well, it's only frightening when you don't do anything or understand what it is." The response is often, "Oh my goodness, the change is so great. I have no way of dealing with it. And it looks like it's going to cause all these very serious problems."

▲ **That is why people tune out, because it just feels overwhelming.**
But in fact we have lived in all the different ecosystems of the globe. We have successfully lived in most of them, including the high arctic. So it's just that we have to do it differently. It's not that we can't persist. We have to do it differently, and it has to be consistent with what's available, what can be sustained and at the same time doesn't increase the risk for the loss of those things that we need, such as water and jobs and maintaining the carbon and food and biodiversity. So it's more of a challenge. People should feel challenged and encouraged to act, to learn, to innovate, to think. Sure, it's a huge challenge for all of us. But that's if we want to stay the same.

On the other hand, it's a tremendous, as they say, teaching moment. The greatest teaching moment of all time, where I hope that we will learn how to live in the landscape and not on the landscape. In the forest, in the broadest sense, and not on the forest. Because that's where we came from. Some areas are lucky. In Canada, we can move around, we can switch to that or whatever, but can you imagine the people living in India? There's no place to go. If the water isn't there or your island drowns because the sea level's rising. Or if there's no more forests, there's no more wood to heat your house or cook your meals. What are you going to do? So I think this is the great teaching moment, the great challenge, and the great opportunity is to learn to live in the land and not on the land.

▲ **So what can we do to improve things?**
The first thing is to become informed. Don't be misled by failed, weakly veiled misinformation. Go to the Intergovernmental Panel of Climate Change (IPCC) website and read what's been approved by governments. The greatest scientific effort of all time is summarized there. You can learn a whole bunch of things and if you read the technical summaries for policy makers, you don't have to read the hard science. You can see what's going on, and what's in there is pretty amazing stuff. This is the kind of stuff that makes people worry and in recent conversations and meetings that's an understatement of what's actually going on. So believe it. Because if you don't believe it, you're an ostrich with your head in the sand. So become informed, number one.

Then number two—look at two levels at which you can take action. Look at your own life and what you do because this consequence of increasing CO_2 is a consequence of our cumulative human activity. It's what you and I want and do. And then work with those groups from the highest levels in terms of government and international organizations to your community groups and municipalities, to change the way you interact with the landscape.

One of the things that I encourage everybody to think about is how you plan the use of the land because the land is where the trees and forests grow, where all those values are and if we pave it, or turn it to things otherwise or abuse it, then those functions don't exist. So land-use planning, very important.

You know it may be that humans have taken enough now from the land and from the seas. It's time to try to live within what we've worked with and taken for our needs in a much more efficient way. So work at the big level in terms of policies that are consistent with basic ecological principles and the risks, and work at the individual level—what you can do. And everybody knows what that is. Hydro tells you to use less power. There's no magic about that. Use less energy. That absolutely has to be done. Demand less from the world.

Think of the quality of your life rather than the quantity. Think about everything that you do and say, "Do I really need to do that? Does it really make me happy, or am I doing it because I'm pressured by my peers or by some advertising or something? What really makes me happy?" If it really makes me happy to have that nice little wood lot to go into in the evening and have a walk with my dog or whatever, then why would I ever support or not oppose the cutting down of all of those trees for giant condos? Why would I do that?

If we've already converted much of the large urban areas for urban purposes and all the large natural values are gone, well let's put more people in there in a way that they live more happily, and they often want to do that. They don't want to be defused over the landscape, but that's sort of the peer group pressure of modern society.

And there are good examples of developments like Dockside Green in Victoria, which are really revolutionary movements toward having no footprints at all other than the physical one, at least in terms of the place where people live. So there are things we can do.

In all your activities, honour nature, honour natural processes, honour and do not forget about the other species. Many of them have been around a lot longer than we have. We have all ridden the same great boat and great adventure sailing the seas of the Earth. We've all got to this point. And right now it just seems to me that we're tossing them all off the boat, hoping that we'll somehow survive on this boat and it will be a lot better for us until, of course, we fill up the boat, and then we'll start tossing each other off the boat.

As we know, people talk about this. Maybe we're all right in Canada, but what about the rest of the world? Indeed, what about the rest of the world? What will the people do in Bangladesh when half of their country is under water? Where will they go? They'll go elsewhere. Eventually, the elsewhere will be here. So the actions that we take are key to every place on the globe, not just here. The values that we espouse and adopt and the way we live, the example that we provide in British Columbia with respect to our forests or anything else, are examples for the rest of the world. We have an opportunity to be leaders. We have an opportunity to develop the skills to teach our children, to learn ourselves and then to go to other parts of the world and other parts of Canada and other places.

British Columbia is very lucky, it has a tremendous tradition of thought, good science. Yes, we've had our challenges. But in many ways we're far forward of other parts of the world because we still have nature so close to us, and we've thought a lot about it and how to observe it and describe it, and other places in the world don't have that. The fact is that natural ecosystems are going to have to be redeveloped where everybody is. The forests will have to come back to all the cities. We don't have a choice. We'll have to do this to work with the challenges we face with climate change.

If the premier walked across the street this afternoon and said, "I'm putting you in charge of our forest policy," what would you do?

Very interesting question. The answer will be in the form of a biologist's question: what should be the future of the forests? A couple of things. One, I think we all have to recognize that people have a legitimate need to use and be part of forest ecosystems. In some cases, very lightly so; in other cases, very intensely so. So let's not imagine that the solution to this is drawing lines all around the forests and all British Columbians saying nobody goes there; we just let them go. That's not an option for a whole bunch of reasons, one of them being that the society and people are a driver of what goes on in the forests anyway. The other thing is, let's look at the forests as ecosystems, and minimize the risk to them through our activities. In other words, there are two influential factors: one is climate change; the other one is human impacts and disturbance in the forests. Then look toward preparing the forests—it's called adaptation—in such a way that they have the possibility to continue to grow as the rich forests we've had. There are all sorts of ecologically based ways of going about this. In some cases, it's preservation; in other cases, it's maybe restoration, in some cases, it's replanting as fast as we can, but always honouring the forests as ecosystems and the processes and the species that are there and all of the values they provide, not only for us as humans, but for all the species that live here.

▲ How do you feel about trees?

I love trees. I hug trees. I think trees are living beings. I think that every person—and my children did this when they were young—should go out and hug trees. Not in the old green way but in the way of "You are an organism, I am an organism. I can't really understand you, but we're beginning to understand that you are more complex and more profound than we ever thought you were."

And when you go out and take a tree, thank the tree. You need the tree, it's a job, we need to make paper or whatever, you don't waste it. You thank the tree and respect it as another species, or as another creature. And I think fundamentally if we think of it this way, and think about their lives, think about the forests as their homes, like your homes, and your community, I think we would go about this in a very different way.

The Royal British Columbia Museum at www.royalbcmuseum.bc.ca

10
BEN PARFITT
BEETLE MANIAC

Ben Parfitt reported on forestry issues for the *Vancouver Sun* until the day he reported on an issue his editor told him "wasn't a story."

"So I said, 'Well, if it's not a story, then I'd like to market it to someone else.' His words were 'go ahead and flog it.' And I did flog it." Parfitt flogged it to Canada's longest running alternative weekly newspaper, *The Georgia Straight,* and almost immediately after it appeared, "I was no longer working on forestry for the *Sun.*"

Parfitt's non-story—about how a spontaneous "grassroots" campaign on behalf of the logging industry was orchestrated by Burson Marsteller, one of the world's top PR firms—led to him being shifted to a few new beats at the *Sun* before he decided to launch a new career as a freelance writer.

Parfitt had moved to BC from southern Ontario in 1986—and became an instant westerner, even though he'd never been farther west before than Sault Ste. Marie. "I remember distinctly getting off the plane in Richmond and knowing that I was never going to go back."

Soon after arriving in BC, Parfitt was hired by the *Sun* as a general assignment reporter on the night shift. But an interest in natural resources—his father grew up in a mining town and his grandfather was an engineer in the mine—steered him to the forestry beat.

Parfitt is the author of *Forest Follies: Adventures and Misadventures in the Great Canadian Forest* and *Forestopia: A Practical Guide to the New Forest Economy.*

He still writes regularly for the online journal *The Tyee,* but since 2005 he has worked for the Canadian Centre for Policy Alternatives trying to come up with new answers for the problems facing our forests.

I spoke to Parfitt in May 2009 about the little bug that's been taking a huge bite out of the forests and the economy, the dreaded pine beetle.

▲ **Was there a particular event that you covered, something that just really hit you and became influential in your worldview?**
Most definitely. When I really got very interested in forestry issues was actually over the last mountain pine beetle infestation. So the one that preceded the current and record outbreak had many of the same elements in play as the current one. It was just on a smaller scale.

When the logging activity in and around Williams Lake exploded in the early 1980s, there was this cascade of events that was really, really interesting to follow. We saw logging rates jump by 50 percent or more. We saw unprecedented rates of logging farther and farther and farther away from the communities where the mills were located. And it raised all kinds of questions about the direction the industry was going in, what the implications were for the many small communities on the backend of the Chilcotin Plateau and what all this was going to mean for the future. That's when I started to get interested in things because the scale of it just seemed to be so out of keeping with what could be sustained.

In subsequent years, I would spend quite a lot of time on Haida Gwaii, the Queen Charlotte Islands, looking at issues there and seeing that patterns were being repeated in many parts of the province, with smaller outlying areas essentially becoming the feed spot for raw materials for other parts of the province, which then took those raw materials and turned them into products and benefited disproportionately from the riches that were being stripped from someone else's backyard. That beetle outbreak in the Chilcotin in the early 1980s was when I really started to get interested in what was happening with our forests in British Columbia.

Some of the stories that have emerged over the years from this beetle outbreak are just astonishing. Weather radar stations were picking up rain clouds on sunny days and those rain clouds were beetles. Lines stretching miles long of dead black beetle bodies washed up on the shores of Tatla Lake. Hail falling on barn roofs in Alberta that turned out to have wings! This was an outbreak of epic proportions.

What do you think we did wrong with the beetles then, and what are we doing now?
In both cases, the approach we have taken is essentially to say this is a resource that is being wasted by the beetles, and the only choice that we really have is to dramatically increase our logging rates and process as much wood as we can now and get as much value as we can right now from that resource before it's lost forever. Unfortunately, what happens when you do that is you just end up compounding problems.

What we're seeing, or are likely to see, in the next little while is a huge amount of dislocation in communities reliant on the forest industry because we're literally eating away at that natural capital too fast for it to be replenished.

In the process, we're also not being very smart about how we're going about logging. So there's plenty of evidence to suggest that perfectly healthy forests have been coming down in the name of salvaging value from dead forests and that this has been happening for far too long. The net result is that we're going to end up with a vastly depleted land base. I fear that without any money being put away for the rainy day, where we have to catch up and replenish and rehabilitate these lands, the consequence is going to be that the depth of the falldown in logging rates is going to be much deeper than we thought, and it's likely to be far more prolonged than many people thought. So people are going to feel a lot more pain for a lot longer than would have been the case if we'd thought things through differently.

▲ **What should people know about the beetles?**
It is a multifaceted problem. Without doubt, part of the severity of this particular outbreak is because of climate change. We have clearly had an extended period of warmer-than-usual winters, which has allowed the beetle numbers to build and build and build, with the end result being an attack more severe than anything we've seen before. And the way we've managed our forests has also helped to sow the seeds for this current outbreak.

Ironically, we've done a very good job of protecting forests from being destroyed by fire. But in doing that, we've allowed for the conditions to become absolutely perfect for the beetles. So by allowing more and more trees to stand and not be burned by fire, we ended up with a proliferation of older pine trees on the landscape that were perfectly suited to attack by the beetles. So it's both a natural phenomena with a generally warming cycle that accounts for what's going on, but also how we manage our forests.

We need to work more closely with more advance light fires to try to create or recreate on the landscape more of the patchwork that would have existed a century or so ago, when we had much more of an uneven aged structure to our forests. We've got to look at changing the ways we approach managing these lands if we want these lands to be healthy and resilient in the future.

▲ **You've been covering this issue for so long that you've got a pretty unique perspective. You've seen Socreds [Social Credit**

Party], Liberals and NDPers [New Democratic Party] come and go on this. Do you find a massive difference in terms of how one party or the other is handling forestry?

It's a very interesting question you ask because one of the points of debate in the current [2009] provincial election is over how we should be awarding forest land to companies. There's a lot of questioning about the tenure system or the system of licence agreements between the province and the various companies that hold logging rights in our forests. Looking back, I think this sort of cherished notion of appurtenancy, the idea that logging companies would get access to publicly owned forests in exchange for running mills in specific communities, has been out there for a very long time. But when I look at it, I see evidence that the Social Credit Party, the NDP and the Liberals all basically at one point in time paid lip service to that idea, while essentially letting it die.

So mills closed, timber rights were not taken back by the government and awarded to other companies. We're left at the end of the day with the bigger fundamental questions about how we award publicly owned resources to entities and what we expect in return.

In my opinion, that bigger question has not been wrestled with in any substantive way over all the time I've been in BC. I'd really like to see much more thought being given to how we empower communities—and by that I mean rural communities, both First Nations and non-First Nations—to work with the natural resources in the surrounding region.

I don't think that it's a panacea to say that community control will provide the answer. But I tend to believe that if we vest communities both with the responsibilities to manage resources—so they have to incur the cost of managing those resources—and the downstream benefits of managing those resources, they will tend to figure out what's going to work best for them. I don't hold any illusions that mistakes would not be made with large community-held licences, but looking at the current system, I think that it's worth giving communities a shot, just as we've spent the last century or so giving companies the right to make the decisions about what goes on on the land base.

What's your role where you are now?

Where I am now I've spent quite a lot of time analyzing forest trends, analyzing what's been happening on the ground, analyzing what's been happening with science, analyzing what's been happening with industries and government's response to things like the pine beetle, then stepping back and saying, "Okay, given all this, given where we find ourselves right now in terms of our approach to this problem, how might we do things differently?" My work is all about analyzing the current lay of the land and proposing why we might want to think about approaching these problems in a different way because doing so might get us better environmental and economic returns than the present approach.

So if you were in charge of our forests, what would you want to do?

The one thing that I would definitely want to do is to promote much more of a diversity of approaches in terms of how we're dealing with things. We tend to do a few things in the forest industry in British Columbia very well. We know how to log extremely efficiently. We know how to mill and process lumber very efficiently. There's room for those things, but if we do too much of those things, if we replicate the same model over and over and over again, we end up in trouble. I think the beetle is a really good example or backdrop to why a narrowly focused approach gets us into trouble. Not all the forests that have been attacked by the beetle are affected in the same way. So you hear horror stories of an area of forest the size of England having been attacked by the beetle. That's true. An area greater in size than England is probably now affected. But that does not mean that all the trees over all of that land base are dead. Yet our approach to dealing with the forest in that land base is essentially to go in and clear-cut every single tree living or dead and take the product and move it as quickly as possible into mills to be processed into lumber. By doing that, two things happen: one, we end up setting the clock, if you will, back to zero on all the lands that we're logging. So everything is cleared out, now we have to replant it and start it all over again. And, two, ultimately we spit out too much lumber.

One of the ironies in our approach to dealing with the pine beetle outbreak was to encourage companies to log as much as possible. The companies did that and they processed as much lumber as possible. The outcome was that the biggest lumber market for British Columbia, the United States, became flooded with lumber. Now here we are on the after-side of one of the worst housing crashes ever and we've got no market for our lumber and we've got an incredibly denuded landscape to deal with.

So I believe that we need to be looking much more carefully at where we log and why we log, not clear-cutting everywhere, looking at selective systems in many places instead of taking all the trees down, and promoting as much as possible a greater array of forest products that we make from the trees that we do cut down in our forests.

I don't believe that we should denigrate the making of two-by-fours. I've used a lot of two-by-fours recently in a home renovation. We need those products. We produce an awful lot of forest commodities; we don't produce an awful lot of value-added products.

We've got to have a far healthier mix in terms of what we make by way of products in the province. And that mix will encourage more people working in rural communities. It will encourage a resource-based economy that's actually more diversified and therefore less vulnerable to the upswings and downswings that we see in housing markets in places like the United States.

▲ How do you feel about trees?

I love trees. They're so important for so many different things. They're beautiful. We have a great array and variety of trees in British Columbia. It's marvellous to be able to walk through forests in different parts of the province and see that diversity. And trees do so much for us. I've talked an awful lot about the products we get from trees, but they do so much more for us. They're critical to any hope we have of trying to moderate and control our climate. They're vitally important in terms of storing and moderating water flows. The list goes on. They're just an incredibly precious and important resource in British Columbia, and we ought to be doing whatever we can to ensure that that resource remains healthy in the long term.

Canadian Centre for Policy Alternatives at www.policyalternatives.ca

11
PATRICK MOORE
ECO-HERETIC

In environmental circles, Patrick Moore isn't just controversial, he's toxic. To most environmentalists, Moore is considered one of the world's leading greenwashers—someone who uses their personal green sheen to polish a company's eco-image.

An early member and one-time leader of Greenpeace, Moore spent 15 years with the organization before founding his own consulting and communications firm, Greenspirit Strategies Ltd. Since founding Greenspirit, Moore has publicly advocated for fish farming, nuclear energy, chlorine, genetically modified crops and using more wood—positions that would make most environmentalists, well, green around the gills.

Moore's apparent conversion began not long after he left Greenpeace, when he agreed to serve on an advisory panel to the BC government known as the Forest Alliance alongside BC International Woodworkers of America (IWA) leader Jack Munro—who once "joked" that if his members were logging and saw an endangered spotted owl they should shoot it rather than risk the area being declared habitat for endangered species. That was when Greenpeace founder Bob Hunter dubbed Moore "eco-Judas."

But when talking to Moore, he comes across as a true believer who arrived at Damascus after the failure of his fish farm in Winter Harbour. "I saw aquaculture as a sustainable development and that was what I wanted to do." After starting his fish farm he became president of the BC Salmon Farmers Association. Not long after that, most BC environmentalists declared fish farms a threat to wild salmon. "Within three years of leaving Greenpeace, or even less,

I was faced with the task of defending my new industry against attacks from Greenpeace, having been a Greenpeace international director for all those years. So that was kind of a switch. And it put me squarely back into politics of an environmental variety. Then I was appointed to the BC roundtable on the environment and the economy, along with 32 other people in the province, and got really involved in sustainability issues. And then I joined the Forest Alliance of BC and lost half my friends. When I joined, it was for the purpose of helping the forest industry improve its environmental performance."

As we sat in his office near BC Place in downtown Vancouver in December 2008, Moore explained that his roots in forestry run deep.

Forestry was Moore's family business for nearly a century. In the mid-1960s, Moore's dad was president of the BC Truck Loggers Association—a position he held for two terms. Then he became president of the Pacific Loggers Congress, which is the association of associations involved in forestry around the whole Pacific Rim, including New Zealand and Chile. He also founded the sports logging shows at the Pacific National Exhibition in Vancouver.

Not surprisingly, Moore went on to study forestry at UBC. His first book is titled *Green Spirit: Trees Are the Answer*.

What is Greenspirit and how did it come about?

Greenspirit was founded to create a vehicle to develop an alternative environmental platform for sustainability that was based more on science and logic and less on sensationalism, misinformation and fear, as I saw the agenda of much of the environmental movement and my old friends in Greenpeace going. I had seen that when I left in 1986, and that's why I left. They were starting to adopt policies that I could not justify or rationalize in terms of the science education that I had. I have a Ph.D. in ecology.

▲ **Do you still consider yourself an environmentalist?**
Absolutely. Yes. It's just the *ism* part that bothers me. *Environmentalism* is a very different word than *environmentalist*, in the sense that the environmentalism implies that there's some sort of bible or some sort of manifesto, or some sort of litany of truths, which are accepted by all.

As an environmentalist, I don't think you have to accept environmentalism in the sense that the other environmentalists are saying, "Oh, Moore, he's no environmentalist." As if they are the givers of the "environmentalist" label or badge. As if you have to agree with them or you're not an environmentalist, because you don't agree with their environmentalism.

▲ **You were talking about sensationalism, misinformation and fear. What were you referring to?**
The two classic examples where those qualities are very clear—there's a lot of others too—are nuclear energy and genetically modified crops. In forestry, for example, there's the fear of extinction of species. Forestry has never been shown to cause species to go extinct. Species go extinct because of agricultural clearances, which include urbanization. Species go extinct because of introduced species coming in, like rats coming onto an island and exterminating a bird species—that's happened in history. That hasn't got much to do with forestry. And species go extinct because of overhunting and eradication efforts. For example, small pox is thought to be extinct; that was an eradication effort.

Overhunting: the passenger pigeon was extinct, the dodo bird was extinct, the Carolinian parakeet was purposely exterminated by farmers in the US southeast. But there's no species we know of that has ever become extinct from forestry. Yet World Wildlife Fund, Greenpeace, etcetera constantly talk about forestry being responsible. Now it is true that in order to establish a corn farm you have to first clear the trees, and they call that forestry? That's not forestry, that's farming.

So they've given the impression to the general public that we need to be afraid of multinational forestry corporations because they are the main cause of species extinction in the world when there isn't a single shred of evidence of that being the case.

▲ **But with forestry, how do you not lose habitat? If you've got no habitat, how do you keep the species alive?**
By planting new trees. Forestry only harvests about 1/80th of the areas of land every year. In agriculture you harvest all the land, every year and leave it completely bare over the winter, whereas in forestry even after you harvest it, you don't take all the shrubs and roots out. New stuff is springing up in about the next five minutes after you've finished forestry.

Forestry is not as destructive as a really bad wildfire, for example. It's not as destructive as a volcano. It's certainly not as destructive as an ice age. But it is a "temporary disturbance," as ecologists call it.

Ecologists refer to these rapid changes in habitat or ecosystems as disturbances. And ecologists all know that ecosystems are capable of recovering from disturbances without any help from anybody. The only thing that prevents them from recovering is continuous human interference and disturbance, such as fixing the pot holes in a road, plowing the soil every year, planting crops and keeping your driveway pavement there.

As soon as people stop interfering and disturbing, nature starts coming back. It was actually a Greek philosopher, was it Horace? The quote is, "You may drive out Nature with a pitchfork, but she will always hurry back."

Having grown up on the northwest tip of Vancouver Island in the rainforest, I know that if I didn't have my machete and my pruning sheers and my chainsaw, the salal bush and salmonberries would be coming through my back door and into my house and covering the whole thing over. You have to keep trimming it back. But that is an act of interference. The fact is if you leave it alone, if you cut just the forest and go away, it grows back again.

Aren't there areas where complete habitats have been wiped out by clear-cutting?
Where?

I'm assuming in BC?
No. There are no areas where that has occurred. Fire will set the recovery back further than forestry in many cases. If you have a fire that actually burns the soil off, which happens on side hills regularly, then it's a lot longer to recover because the soil has to recover first. But most logging does not destroy the soil, it just cuts the trees and leaves the seeds and roots and everything in the ground to regrow. But no, there is nowhere there has been permanent loss of forests due to forestry. Forestry allows the forest to grow back and in most cases assists by planting the same species that were there in the first place.

Even something like the Bowron Lake clear-cut?
Yeah, you should go see it now. You'd never know where it was. You couldn't find it anymore. It's all grown back into quite big healthy trees. And right now this huge clear-cut that's being caused by the beetle epidemic—which by the way is not a unique occurrence, this has been going on for millions of years. Most people don't realize it, but the average age of the forest in the BC interior is considerably older now than it was when we first started logging there 50 or 60 years ago in a big way. And that's because we've suppressed fire, whereas normally fire would have taken out a lot of those forests when you let them get older, then

they become more susceptible to beetles and the combination of the warmer winters that we've had in the last two decades plus their age—once lodgepole pine gets to be about 100 years old, it starts to decline and becomes more susceptible to infection and beetles. So it's sort of a perfect storm that's occurred there partly as a result of management, partly as a result of climate. And these huge clear-cuts that we're creating now as we're chasing the beetle and trying to use the wood that's dying, rather than cutting green trees, those will all be beautiful second-growth native species forests in 30 or 40 years from now.

We were talking about environmentalism as a form of religion. When I was doing my research before meeting you, one of the stories that caught my eye referred to you as "eco-Judas." What is your main heresy to the environmental movement?

The specific issue that caused Bob Hunter to coin the term *eco-Judas* for me, and Bob was one of the most brilliant wordsmiths who every existed—

It was Bob who gave you the name?

It was Bob. He also said I was "schlepping for the stump makers," which I thought was quite a good turn of phrase as well. Anyway, Bob Hunter was the one who coined the term, and it was purely and only to my agreeing, at Jack Munro's invitation, to join the Forest Alliance of BC as a director. And I was invited by Jack Munro for a number of reasons. First, he had been selected by the CEOs to chair the Forest Alliance as a way of showing solidarity with the working people who were going to be involved in whatever was going to happen as a result of this initiative. So that made sense to bring Jack in.

Jack had never been the type of firebrand—I mean, he swore a lot and pounded his fist against the table—but he was not a snob. He would break bread with the management rather than treat them as the enemy. He knew what his job was: to get a better contract for his people. But he didn't regard the management of the forest industry as being some kind of evil force. So that worked to bring him in.

Also, Jack Munro and my father, Bill Moore, were very close friends for many, many years. My dad was the first, and at the time only, small logger invited to sit on forest-industrial relations, which is the bargaining arm of forest industry management that would bargain directly with the IWA at the time.

I had at that time left Greenpeace three, four years earlier and was now running a salmon farm on Northern Vancouver Island. So I wasn't just a one-dimensional person. And they wanted some people with environmental credentials on this Forest Alliance if for no other reason than to give it credibility, but also they wanted some environmental people on it . . . I was the only one with genuine environmental movement credentials that was willing to help the forest industry improve its environmental performance because the environmental movement saw the whole thing as if it was just an image, PR exercise, whereas it's very clear, if you look at the record, that the practices of the forest industry changed dramatically through the 1990s to where they were nothing like they had been.

What are you proudest of in terms of your work with the Alliance?

The thing I'm proudest of is that we convinced the industry in a general sense to approach this in a positive way, rather than a defensive way, to engage rather than to fight, to get involved. And they did get involved. And the chief foresters got involved.

By year two of the Forest Alliance, the chief foresters of all the major companies were working together to work with the parks and wilderness and forest practices people in government, instead of just being against everything that was being proposed. And we did that. Nobody else could do that.

▲ **What accomplishments are you proudest of since Greenpeace?**
Since I left Greenpeace, I'm proudest that I was one of the founders and developers of the salmon aquaculture industry in British Columbia. I think it and aquaculture in general are very important. I simply reject the story that the fish farms are killing wild salmon. I do not believe that is true. I believe it is a fabrication.

▲ **If somebody put you in charge of our forests today, what would you do?**
To tell you the truth, I don't know that I would do an awful lot different than what is being done because I was part of the process of changing the forest practices to what they are like now, rather than what they were like before. I think I would certainly, as the government has kind of indicated, state that the park-creation era is now over, that the goalpost can't just keep moving forever. First they say they want 12 percent, and then 15, then 20, then 50, then where does that end? It is okay in British Columbia to use the base of forests that has been now accepted with the parks and wilderness having been created and more than doubled. We now have at least 15 percent of British Columbia in formal protection, but there's also about 60 percent of British Columbia's forests that is unsuitable for forestry and therefore will never be used for forestry in any foreseeable future. And that is the high alpine forest, the too-steep forest, the scrub forest of the west coast, where it's just nothing but little wee short stunted trees in boggy landscapes.

The thing that always upset me about this question of which areas do we preserve and which areas do we log is that the environmental community and the anti-forestry community generally seem to assume that the best areas for growing trees were also the most important areas to preserve in perpetuity. In other words, there's no compromise here. Whereas I know that in fact many of the areas that are not suitable for forestry are also extremely important from a biodiversity point of view and in many cases can preserve the values that exist on the areas that are useful for forestry.

Forestry is like agriculture in the sense that you can't farm everywhere. You can only farm where it's suitable to farm. And that is usually flat places. At least forestry can be done on landscapes that are hillier and steeper than you can usually do agriculture, unless you get into terracing and all that like they do in the tropics, but that wouldn't be particularly useful here.

So I've always lamented the fact that this always comes down to such hardcore politics, when in fact there's plenty of opportunities to say, "We've got this huge area here that is not suitable for forestry, that is quite capable of preserving these values, so let the forestry happen over here." But it's never been that way. Will the battles of Clayoquot and the Great Bear Rainforest never end? Is this an eternal struggle that will go on for the rest of time?

Surely at some point we could decide that the land-use plan is suitable and sufficient for the time being, and that we have a system in place that will both allow a viable and sustainable forest industry and at the same time allow for the protection of those values in biology and ecology and biodiversity that we want to protect.

So I would foster continued roundtable, multistakeholder processes. It seems to me that we've lost that. We don't have a lot of processes going on because a lot of it's been resolved. Many of the jurisdictions in BC have been signed off. But Clayoquot and the Great Bear Rainforest remain sort of ongoing sores in the dialogue about how we should actually manage the resources and zone the land. In my estimation it all comes down to land-use planning and zonation—that in this area you can do this, this and this, and in this area you can't do anything. We need to finish resolving those subjects, is what I would say. That's what I would work on.

As far as forest practices themselves go, I don't think there's a lot more to do there. I think we have done a very good job in BC of adopting world-class sustainable forestry methods and practices that will see our forest industry move into the future. There are always going to be natural events, like the bark beetle and climate change, etcetera, that are going to interfere with our perfect idea of how things should be managed. The only constant is change. Nature never operates exactly according to a clock or anything. And so we're going to have to deal with the bumps and hiccups of the natural world, but at the same time we have a large enough land base and we know how to grow trees and we're doing forestry in a very natural way in British Columbia, more so than in almost anywhere else in the world. We are reforesting with native species and fostering a natural-like forest to come back in as the second-growth or third-growth forest.

I love that between Campbell River and Kelsey Bay there is a third-growth forest at the crossing of the Amor de Cosmos Creek. The creek was named after a premier of British Columbia. If you look there now, you will see a third-growth forest growing back, as the second-growth has already been logged again. And you could never tell that from this young native forest. Though there was planting there, other species fill in by seed and you end up with a forest that looks very natural.

▲ **How do you feel about trees?**
I love trees. My only book is subtitled *Trees Are the Answer*. I believe that trees are the answer to a lot of questions about our future here on this Earth, ranging from the most practical, of

what we should build our house with, to the most aesthetic, about how to make the world more beautiful and green. And in between that comes how to pull carbon out of the atmosphere. Trees are the most powerful carbon pump we know. They are also the best solar panel that we have for producing energy and material. Trees provide wood, which is the most abundant renewable energy source in the world and also the most abundant building material in the world, as well as the byproduct of paper, which is used for printing, packaging and sanitation. The simple products of the forest are some of the most important for wide-ranging uses. Then there is the fostering of biodiversity, wildlife, etcetera. Growing forests is always better than planting farm crops, annual farm crops. It you want to have birds and animals on the landscape, grow forests.

The other is cleaning the air, cleaning the water and building healthy soils. There's nothing better to do that than trees. So the balance that needs to be struck is ranging all the way from the beauty of nature and the beauty of trees and forests, which is what your wilderness parks are for, all the way through to the intensive management of forests for material production, while at the same time you're providing the benefit of nature conservation, biodiversity conservation, clean air, filtering the air, filtering the water and building healthy soils. So trees are the answer.

Greenspirit Strategies Ltd. at www.greenspirit.com

12
KALLE LASN
ADBUSTER-IN-CHIEF

In the late 1980s, the BC forest industry responded to progressively more visible and vocal anti-logging protesters with a TV advertising campaign designed to assure viewers/voters that there was nothing wrong with the way logging was managed in BC and that despite the fact that we had a clear-cut that could be seen from outer space, we would always have forests. The campaign was called Forests Forever.

Freelance filmmaker Kalle Lasn was so outraged by the ads that he teamed up with some friends to create a response—and in 1989 he produced a claymation-style ad featuring an ancient tree talking to a sapling, explaining that "a tree farm is not a forest." Lasn hoped his ads would counter the pro-forestry propaganda. He never found out.

Much to Lasn's shock, Canadian networks refused to show his anti-forestry commercial, arguing that they didn't show "advocacy ads"—a surreal claim considering there was no other way to label the Forests Forever campaign, unless they went with the classic definition of "propaganda." And isn't an ad for diamonds or Porsches kind of advocating consumerism?

Lasn's outrage at discovering that all Canadians didn't have equal access to Canada's airwaves fuelled a lawsuit; created Canada's most internationally successful magazine, *Adbusters* (which publishes more than 100,000 copies of every bimonthly issue); launched one of the most unique protest movements in modern history; popularized the concept of "culture jamming"; and celebrated the ultimate unholiday for consumers worldwide, Buy Nothing Day.

Ever since his anti-forestry ad was chopped down, Lasn has been battling bimonthly for our "mental environment" via *Adbusters*, which has featured such satirical print ads as the Joe Chemo campaign (featuring a bedridden smoking camel).

But Lasn's never forgotten the court case that launched it all—a case he finally won in April 2009 after spending more than $20,000 to fight for the rights of citizens to express themselves on Canada's airwaves. It's a fight he hopes to bring to other countries around the world, including the US, where almost every major TV station has refused his advertisements for Buy Nothing Day.

I met Lasn at his office in Vancouver a few days after his big victory.

▲ **How are you feeling?**
That was an incredible moment. After 20 years of being knocked back again and again and again until I was punch drunk suddenly, wow, the court gave us a victory.

▲ **Why did it take so long?**
I don't really know. It could have been the judge, who read a recent issue of *Adbusters,* for all I know. I have a feeling that it's got something to do with a change in *Zeitgeist.* When we first launched this case 20 years ago, yeah, it was a bit of a joke. People didn't give a damn. Who cares whether you can buy or you cannot buy on television? But the world has become a much more dangerous place over the last 20 years. All of a sudden we're faced with catastrophic climate change, and there is a possibility that things might tip over in a way that is very, very damaging for future generations, maybe even for hundreds or thousands of years to come. There's that ominous thing hanging over us. Physiologically we're caught in an epidemic of mental illness where depression and anxiety attacks are

going up exponentially. And of course politically we all know we're in this never-ending war against terror. So ecologically, psychologically and politically we're hitting the wall. All of a sudden people are wondering whether they're really getting the information they need to deal with the future. And all of a sudden the idea that the forest industry can go on TV and buy $6 million worth of airtime telling us that we have "forests forever" and a citizen of Canada cannot go on that same TV station and buy 30 seconds of airtime to tell the other side of the story—that imbalance of who has the power to disseminate information becomes important. I think the judge may have got wind of this *Zeitgeist* shift.

I know how this case started and I really want you to tell the story.
Adbusters Media Foundation was born out of a fight that we had with the forest industry in British Columbia. In 1989, they had this multimillion-dollar campaign with ads on TV, in bus shelters, full-page ads in newspapers, and they wanted to make British Columbians feel good about what they were doing. So this campaign told us that they were doing a wonderful job in our forests and that we have nothing to worry about. We have "forests forever," was their campaign slogan.

A few of us environmentalists and filmmakers came up with a 30-second spot that tried to tell the other side of the story. And then something happened, which was for me a defining moment in my life and gave birth also to the Adbusters Media Foundation. When we came at the same stations that were selling millions of dollars of airtime to the forest industry with our 30-second mind bomb, they refused to sell us any airtime. That was when I realized that there really is no democracy on the commercial airways, that the big-time sponsors would have their say and the dissident voices would be kept out. That was the moment that gave birth to everything that we've done since.

What was it that outraged you so much about the Forests Forever campaign?

Well, it was just a fucking lie. They were telling us that we have "forests forever," and the old-growth forests at that time—as far as I remember—were already down to 17 percent of what they used to be. The annual allowable cut was well above sustainable levels. They were doing a horrible job managing our forests. And this $6 million campaign was just a PR smokescreen for them to continue their carnage.

The people who were really in the know about what was happening in our forests were totally on our side. And when we started talking back against the forest industry and had this fight with the CBC and other private broadcasters, we had a hell of a lot of support from all kinds of forestry people and environmentalists who wanted this carnage to stop.

What was the official explanation you were given for not being allowed to run your campaign?

The private broadcasters basically said, "This is not an ad. You're not selling anything. We're not going to sell you any airtime."

The CBC was a little more nuanced. They pointed to some article 3 of their advertising policy which said that advocacy advertising is not allowed on the air. Their rationale was that if advocacy advertising was allowed on the air, then the people with the bucks, the people with the money, the moneyed interests would be able to control the public agenda.

So the forest industry, which apparently has no money, was allowed to purchase ads—

And did what's obviously an advocacy ad.

▲ **And that's the nuance I'm not getting. How did they justify with a straight face that the forestry ad was not an advocacy ad, but your ad was?**

It was really interesting what actually happened there. Initially they told us to "go away, we're not going to sell you any airtime." And then we started going on radio shows and writing letters to the editor and creating a bit of a ruckus. We came up with a newsletter, which eventually turned into *Adbusters* magazine, and they were suddenly confronted with exactly the question that you asked—how come the forest industry is allowed to buy time, but this group of people can't? And eventually what they did was they pulled the forest industry ads. They said you're also advocacy, we're not going to sell you any airtime either.

▲ **What was your environmental involvement before doing the film, before going against the Forest Forever campaign?**

In the 1970s and 1980s, bit by bit I really got scared about what was happening to the planet. I suddenly realized that this human experiment of ours on planet Earth, that we were going through some sort of existential moment where our own impact on the planet was starting to be catastrophic and unless we found ways of dealing with that we would really hit the wall in ways unimaginable. So I slowly became a greener and greener kind of guy. When we went against the forest industry and couldn't air our environmental ad on commercial television, that was sort of the moment where I tipped over the balance and became a rabid greeny and started fighting the fight.

▲ *Adbusters* **in the early going especially was very environmental. Can you talk about where your passion shifted?**

My passion didn't really shift. When we launched the first few issues of *Adbusters* in 1989, we were one of the few publications, one of the few voices, that was actually boldly and radically talking about environmental issues. But then, even a couple of years later, by the early 1990s, after that first Rio conference, all of a sudden some surveys were showing that 70 or 80 percent of all people now saw themselves as environmentalists. And I started picking up the *Vancouver Sun* on the weekend and seeing articles by people that were actually better written than the writers we were able to recruit for *Adbusters*. So bit by bit the green voice, the environmental voice, rose up and we at *Adbusters* morphed into mental environmentalists—we're still very passionate about the environmental cause. We wanted to launch a mental environmental movement, which we saw was the other side of the physical environmental movement coin. That was a new field, something that nobody was talking about, and we felt like we had a unique role to play and a unique voice to offer.

Is that why you think you caught on worldwide?
Yeah. I think that we caught on worldwide because we were one of the first voices to boldly speak back against consumerism. This was in 1989 and the early 1990s. Anybody who started talking about the fact that we have to live more frugally, cutting back on our consumption and that consumption has a dark side to it that we haven't woken up to yet, were dismissed as kooks. So when we started talking that way and launched Buy Nothing Day, which became a worldwide campaign in the mid-1990s when we put it on the internet, I think we became a worldwide movement. But the other reason is that, right from the start, we said that we were tired of all the old movements. We're tired of feminism, we're tired of environmentalism, we're tired of the old political lefty discourse. We felt that we the people were losing the ability to generate our own cool, that we were losing the power to generate our own culture from the bottom up. More and more of our culture was being spoon-fed to us top down by advertising agencies and corporations that wanted to sell us cars and whatnot.

Companies like Nike were creating a kind of top-down corporate cool that somehow teenagers were buying into. So we said, "Okay, this has to stop." Once we the people lose the ability to generate our own cool, our own culture from the bottom up, then we've lost everything. Then somebody else is basically doing the culture generation, doing the talking and giving us what's cool. Once we lose the ability to create our own cool, then something is severely amiss. We launched what we called the culture jamming movement. Our aim was to get consumer culture to bite the dust and then to build up a new culture on the ashes of this consumer culture. And we called this "culture jamming." And this movement, even today, is one of the hubs of global activism.

If somebody came up to you tomorrow and put you in charge of BC's forests, what would you want to do?

I'd probably kick out all the old guys first. That would be my first move. I'd just get rid of them and start changing the forestry culture that we've had here. I don't think there is any quick fix, but I think we have to change the culture and change the way we think about our forests and change our strategies. I would pull off a kind of Obama-like slow shift of changing the culture.

I would get the best people in British Columbia together—maybe 10 of them or 20 of them, whatever—and I would say, "All right, let's design a new forestry culture."

How do you feel about trees?

I have about 150 of them on my place in Aldergrove, and I prune them, and I eat their fruit, and I have one tree there that I've designated to be my mother, who died recently, and I talk to that tree every now and again. I have a whole grove of walnut trees—and I guess it sounds kind of petty—but I must admit I do love trees.

Adbusters Media Foundation at www.adbusters.org

13

TZEPORAH BERMAN
ECO-ROCK STAR

In October 2008, I attended a party in Vancouver celebrating ForestEthics' astonishing victory in convincing the BC government to preserve 2.2 million hectares of habitat for the world's only mountain caribou. The key players in the fight were introduced to the audience and applauded. Everyone else in the organization was identified by their official titles. Tzeporah Berman, the group's cofounder, was introduced as "our rock star."

Berman has been on the front lines of the environmental movement for more than a decade. But she achieved Suzuki-esque status when she was chosen as one of the featured players (along with the sainted Dr. David) in Hollywood's cinematic plea to save the planet, *The 11th Hour*. Thanks to her turn in the spotlight, Berman scored a surreal official audience with the icon of our age, Paris Hilton, which turned into the kind of environmentally friendly celebrity zoo the world hadn't seen since Brigitte Bardot protested the clubbing of seal pups and helped ensure that, as Berman says, "green is the new black."

After seeing a slideshow about BC's Carmanah Valley in university, Berman donned a backpack, flew to BC in the summer of 1991, stayed at the youth hostel in Jericho Beach and volunteered with the Western Canada Wilderness Committee. She was taken up to the Carmanah and spent the summer as a cook, clearing trails and holding the sound boom for a documentary about the rainforest. She came back the next summer and a Sitka spruce grove she'd fallen in love with in the Walbran Valley had vanished and "a sense of urgency and outrage hit me that I'd never felt before." The next day she hopped in a van to Clayoquot Sound and helped the protesters there. After many of BC's top environmentalists were arrested and banned from returning to Clayoquot, Berman found herself one of the "senior" protesters on site. That led to Berman spending seven years with Greenpeace International and Greenpeace Canada. While working with Greenpeace, Berman decided to try to hit logging companies in their wallets and helped found ForestEthics—an organization that works with and/or bullies businesses into better environmental practices.

One of ForestEthics' biggest early success stories was convincing Victoria's Secret—yes, those would be the lingerie people, as well as the top catalogue manufacturer in the world—to stop using old-growth paper for their catalogues. The Victoria's Dirty Secret campaign featured a lingerie model wielding a chainsaw.

ForestEthics is currently fighting to ban junk mail in the US "because junk mail in the US is 100 million trees per year," says Berman. "We think if we're able to reduce the stream of junk mail by just 25 per cent, it will be the equivalent of taking a million cars off the road."

I met Berman at the ForestEthics office in downtown Vancouver not long before she left for Bali in December 2007 to take on Canada's environment minister on the issue of climate change.

Since we spoke, Berman has left ForestEthics to help found PowerUP Canada to fight climate change—and it was clear during our talk that her concerns were already shifting from saving the forests from chainsaws to saving the planet from global warming.

In our interview, the uncompromising activist talked about the challenges of making compromises, the power of powersuits, staying hopeful even when you're worrying about doomsday, the sticky situation in the tar sands and the value of celebrities.

▲ **How did you get involved with *The 11th Hour*?**

I was at an amazing annual conference in Marin [California] called Bioneers. I was speaking there on my journey from blockades to boycotts, and after my speech I was standing outside and there were some people talking and I could hear them saying, "Oh my god, David Suzuki is talking about this, and we've got Paul Hawken talking about this, and Gorbachev saying this, but we don't have anybody in this whole film who can tell us about the state of the world's forests."

I'm not shy, so I turned around to them and said, "I think you need to look at the world's resources and data that's showing that 80 percent of the world's intact forests are already gone and there are only three countries left in the world with enough forests to maintain biodiversity and ecosystem services. And that's Canada, Russia and Brazil."

Five of these really intense LA types, you know, Hollywood types, are all chain-smoking and they kind of turned around to me en masse and said, "Who are you?"

And it turned out they were all the directors and producers of *The 11th Hour*. Two weeks later they flew me to Hollywood and interviewed me on the state of the world's forests for the film. They interviewed me for hours. They interviewed 80 people for that film, and I think there were 40 people who ended up being in the film, and I was one of them.

▲ **You mentioned at *The 11th Hour* premiere in Vancouver that you got more press for saying hi to Paris Hilton than you had for what seems like a lifetime of environmental activism. How? Why?**

The producers called me in July and said, "We screened the film at Cannes and it got a standing ovation and Warner Brothers International offered to buy it and if they do that, it looks like the premiere will be really soon—like within the next month. Will you come to LA for the premiere?"

I was one of the only Canadians, one of the only women in the film, and I had really connected with Leila and Nadia, the producers, and so I said, "Okay, I'm going to try." And I thought, "When is this going to happen again? I'm in a Hollywood film. I'm going to go to LA to meet Leonardo DiCaprio. I'm going to do it up. I'm going to do it right."

So I called a bunch of friends to ask, "Do you know anyone who knows anything about PR in LA because I'm saying this about Canada in this film and maybe we can use it to help on the issues."

We ended up hiring a publicist and we held this little event the evening after the premiere for the organization that I founded, ForestEthics. And two things happened. One is that the press went crazy—Canadian press went hard with the "young Canadian activist walks the red carpet." Or maybe they didn't say "young," maybe I'm just flattering myself now, but they said, "Canadian activist walks the red carpet."

The Canadian press went crazy, and it gave me an incredible opportunity to talk about what's happening in Canada's forests and why Canada's forests are so important and also about global warming. And the next night after the premiere, at the ForestEthics event, in walks Paris Hilton and Adrian Grenier from *Entourage* and a number of other celebrities who had heard about ForestEthics and our event and *The 11th Hour* and they decided to come. When Paris Hilton walks into the room, so do a hundred paparazzi trailing her.

▲ She really has her own entourage like that?

Oh yeah. Her car pulls up and all of a sudden there are other cars screeching, and there are police and people are putting up barriers and some woman comes up to me and says, "I'm Paris Hilton's blah blah blah. She's here, she'd like to be briefed on who ForestEthics is and what you do, and do you want your picture taken with her?" And I had like 30 seconds to go, "Mmm, okay."

So I walked over to her and I said, "This is what we do and this is why I think it's important and I'm glad you're here."

And she said, "Great."

And we stood together. And the photos are hilarious because she's doing this fabulous little sexy pose and I look like a deer in the headlights.

It all happened so fast and 100 flashbulbs are going off. And people are screaming at you, "What are you wearing?" and "Why are you here?" Really, it was a zoo. It was totally crazy. I was probably with her for all of two minutes.

I briefed her, had a chat with her, thanked her for coming, talked to her about ForestEthics and how important Canada's forests are, and then we helped sneak her out the back door and she got a copy of the film so she could see it at home. It was an amazing thing. So she leaves and then all of a sudden 100 people go running out, stop traffic and screech in their cars to chase after her because she's going home. The whole celebrity thing is very bizarre.

That two minutes went on the front page of almost every newspaper in Canada, in full colour, above the fold. And my phone didn't stop ringing for about two weeks. I did more interviews back to back with every press outlet—and strange press outlets, like *Entertainment Tonight Canada*. I'm used to doing news interviews and this was a lot of celebrity press. And it was great because they'd all say, "What's she like?" and "Have you met Leonardo DiCaprio?" And I'd say, "You know, they're very concerned because—" and just start talking about the issues. So it was a great opportunity.

▲ **Paul Watson [founder and president of the Sea Shepherd Conservation Society] talked to me years ago about bringing**

Brigitte Bardot out and how that was the only way to get attention for the seal hunt. I'd actually assumed that you'd called Paris Hilton up just to get attention.

No, she just showed up at our event. What a strange thing that kind of celebrity-ness is. I got this wave of enthusiasm and interest in the issues on my Facebook page and all over the place from 14-year-old girls. My niece was going crazy; all of a sudden she really wanted to know about the issues and she wanted to understand them because they heard about it because they saw a picture of me with Paris Hilton.

It got big play, not just in Canada, but all over the world—Paris Hilton going green. We probably did as much for her image as she did for the issue, which is fine. And it really did get play all over the world. Then about two weeks later she was photographed walking along Queen Street in Toronto wearing a T-shirt that said, "What if this was the last tree?" and it had a little sapling on it, or something. So she's starting to think about these issues.

▲ **What's your next Canadian campaign?**

There's no question in my mind that what's happening in the tar sands is the biggest environmental issue and, I think, actually will become a massive constitutional issue in Canada. Canadians have such a high environmental awareness now. You see everyone from municipalities to companies to individuals struggling to reduce their greenhouse gas emissions, and so few people know that we, our country, is in the process of creating the largest single fossil fuel development project in the world. In fact, it is now considered the largest industrial project of any kind in the world. The scale is staggering.

▲ **I get the impression that people know the scale, but they don't know the cost.**

Right. And the cost is that over the next 10 years it will contribute massively to increasing our emissions at a time when we've committed to the world under Kyoto to reduce our emissions. So even if we all change our lightbulbs and buy a Prius and municipalities try to go carbon neutral, it won't matter because the tar sands negate everything. Under the Harper government, under the new "intensity targets," they can just continue to be pumping out fossil fuels and emissions at an incredible, incredible rate.

Did you know there are now 51 square kilometres of toxic lakes—you can see them on Google Earth—51 square kilometres of toxic lakes. Those are the tailing ponds from the tar sands. There are people employed just to rake the dead birds off the top of the tailing ponds. It's the biggest toxic nightmare almost anywhere in the world and the tar sands are exempt from almost all air and water regulations. So in my mind there's no question that this is the issue we as Canadians have to be addressing.

One of the untold stories of the tar sands is that incredible, lush boreal forests, original forests, are being destroyed for the tar sands. The tar sands lay under the boreal forest. So there is a massive logging project going on to get at that oil as well.

That's funny because in my imagination that's a barren area. That's what people think of it—as sand. No, it's stunning intact forests that are home to billions and billions of water fowl and some of the most important areas of nursery and breeding ground for North America's birds and some of the most beautiful and incredibly important freshwater areas. The boreal as a whole has more fresh water than any ecosystem on Earth. These areas in Alberta are absolutely stunning. So at a time when we should be figuring out not only how to stem global warming, but how we're going to adapt to a changing climate, we're destroying the healthy, functioning ecosystems that we have that are our natural air and water filtration systems. These forests hold the key to our being able to function in the world in a changing climate. It's insanity.

▲ **One of the things I find fascinating is ForestEthics' work with the companies. Did you have any resistance to that? Did you have any people say, "No, no, you can't work with the companies—all logging is bad."**

Oh, for sure. And it's not easy because when you look at what's left globally, 80 percent of the world's intact forests are already gone. On a purely ecological basis, I don't think we should be logging old-growth at all. Even if we stopped logging old-growth today, we can't be sure that the ecosystem remnants that we have left will survive in any form of where they are today because they've already been so heavily devastated.

So there are still, today, a lot of people who criticize us. Well, not a lot, actually. A couple of key armchair activists who criticize us because they think we should be taking harder stances and saying absolutely no logging. But in my mind, if we just create little parks, and we're not working to change the way the economy works and to help create sustainable jobs and a conservation-based economy, then those parks will just be reopened to logging in 10 years. So what we decided to do in the Great Bear was protect all the intact valleys.

There were 359 valleys originally on the coast, and when we started the campaign, there were 69 left. All 69 were slated to be rotary logged in the next five years. So we said, "All 69. We want all 69. We want all the intact valleys." But that still leaves tens of millions of hectares of forests that are very fragmented and already have roads in them, but it's old-growth.

We said, "Leave the pristine areas alone. And use that as your test case to see whether you can do good ecosystem-based management and actually derive some economic benefit while maintaining enough of the structure and composition of the forest to have a healthy ecosystem, not to do clear-cutting, but to do something different."

And there were people who criticized us, and that was difficult sitting with those companies. But what do you have to negotiate if you're sitting across the table with someone saying, "Actually, what I really want you to do is go out of business, and I want all logging to stop and do you think you could do that?" There is nowhere to go with that.

Already we were pushing for so much. Nobody had ever negotiated for more than one valley at once in this province. We came in and said, "We want the west coast of the country with everything that's left intact."

At the time, we got criticized by other environmental groups because we were asking for too much. I had some people who remain unnamed at this point come in and say, "You're making the environmental movement look hysterical and crazy. You can't ask for that." And then years later when we signed the Great Bear agreement to protect 5 million acres [2 million hectares] of rainforest, every single one of those damn valleys—we got criticized for supporting any logging at all in old-growth.

Where you draw the line, I think, is one of those most difficult parts of this work.

How do you stay hopeful? How do you think people can stay hopeful?

For me, the fact that *Vanity Fair* is doing a green issue and Paris Hilton is wearing tree T-shirts is hopeful. Our job as environmentalists used to be to kick open the door to the issue to get people to notice. Well, we don't have to do that anymore. People know. So now we can focus on pushing solutions, on developing solutions, on working with decision makers to identify the appropriate legislation, etcetera.

Our job is different today and in some respects it's easier, and I think that's hopeful. Barbara Kingsolver once said that, "as a parent, hope is the only moral choice," and I really feel that way.

My children remind me to have fun and play, and they also remind me every day how I can't do anything else, I have to do this.

And success keeps me going. In British Columbia, ForestEthics just helped to ensure that more than 90 percent of the mountain caribou's core winter habitat is protected in the inland temperate rainforest and that means about 400,000 hectares of old-growth forest was protected last month. And that's big. You know, 400,000 hectares is bigger than all of Clayoquot Sound. That feels good to know that that's protected forever. The fact that I can do that, and we can do so much more because if there was ever a moment to be working on these issues, when you have the attention of decision makers, this is our moment. That's pretty hopeful.

▲ You've been saying that this is the tipping point for us. Anything you'd advise people to do?

Organize. Don't sit there and wait for someone else to organize something that you're going to be a part of. We're seeing climate leadership in Europe because groups of 5 and 10 people in their communities are going to protest at city hall for a carbon neutral city, or they're marching. The folks in Vermont marched 10 days. Ten of them started and by the time they got to Burlington there were 1,000 of them. And now Vermont is going to be one of the first states to have a low-carbon fuel standard.

People need to organize. We need to hit the streets. We need to sign up and support environmental groups that are working on these issues, like ForestEthics and so many other environmental organizations, because I think that we're responsible today not only for what we're doing and what we're buying and whether we're buying the right lightbulbs—we're responsible not only for what we're doing, but for what we don't do. I think that if we look back in 10 years and know that we didn't have the courage to stand with a placard in front of the legislature at a time when our very climate was threatened, then we'll regret it.

▲ **If somebody put you in charge of the forests today, what would you do?**
If someone put me in charge of the forest, I would protect all the old-growth that we have left. I would immediately ban the export of minimally processed products and raw logs because I think we can support as many people, if not more, working in our forest today if we just get more jobs for every tree cut.

I think we need greater value-added manufacturing to ensure that we still have healthy communities and we have healthy forests. So I would protect what's left. I would keep more of it at home. Honestly, to protect our forests—as great a threat as industrial logging is—probably a bigger threat now is global warming. So I think for all issues, if I could pass my magic wand, I would ensure a strong carbon tax and emissions reductions across the board at a provincial and a national level and set us on a pace for a carbon-neutral Canada.

▲ **How do you feel about trees?**
Being in the forest is what makes me strong. For me, trees are life. When I have those moments when I need to be strong, I picture those Sitka spruce trees in the Carmanah Valley and that's what I try to emulate.

ForestEthics at www.forestethics.org
PowerUP Canada at www.powerupcanada.ca

14

ALAN DRENGSON AND DUNCAN TAYLOR

WILD FORESTERS

Alan Drengson and Duncan Taylor are environmental studies professors at the University of Victoria who have teamed up to find how forests should work.

Alan Drengson is a professor emeritus of philosophy and helped found UVic's environmental studies faculty. He grew up in BC logging communities and spent time playing in the forests and later working summer jobs in mills, doing "everything from off bearing a head saw, which is incredible dangerous, noisy work; feeding the hog the big slams of bark that they didn't want to use in those days so they all went into the chipper; to straightening out the enormous timbers that came off the headsaw so that they could be sorted and sent back to the resaw and then that whole chain fed the green chain." He's the founder of two journals: *The Trumpeter: Journal of Ecosophy* and *Ecoforestry*.

Duncan Taylor was UVic's first full-time faculty appointment in environmental studies. His earliest experiences with trees included raising apples in a family orchard in Ontario.

Their first book together—*Ecoforestry: The Art and Science of Sustainable Forest Use*—collected essays exploring "radical" and "holistic" approaches to forest stewardship and philosophies and experiments in "ecologically responsible forestry." Their inspiration for the book was legendary Vancouver Island logger Merve Wilkinson—known worldwide for the way he manages logging on his own sustainable forest, Wildwood.

Their 2009 book, *Wild Foresting: Practicing Nature's Wisdom,* explores the importance of forests economically, socially and spiritually and features 40 essays examining concepts ranging from ecophobia to ecopsychology.

We spoke in October 2008 at Drengson's kitchen table in Victoria, BC, in a house surrounded by a diverse collection of beautiful trees.

How did you both become passionate about forestry?

DT: My first real interest in trees began when I was a teenager. We acquired a farm near Georgian Bay in Ontario and put in 2,000 apple trees—McIntosh, Northern Spy and Delicious. I was the chief orchards person, which got me interested in not only trees, but also in organics and what we could do to have some of the acreage put into organic apples.

I then came out to British Columbia, and my older brother was a paramedic working at the MacMillan Bloedel camp north of Golden River. I'd go up there for visits. One day one of the fallers was going around with a camera. He was taking pictures of trees. And I said, "Why are you taking pictures of trees?"

He said, "Don't get me wrong, I love trees. But I'm going to be cutting these ones down. Somebody's got to do this."

And that got me very interested in the values of protecting forests versus jobs. I later went on to investigate and write about sustainable forest communities. During that period I met Alan and we began to collaborate on ways to protect intact forest ecosystems and at the same time protect communities that derive their sustenance from those forest ecosystems.

AD: Yeah, vibrant communities and long-term sustainable forests. My passion for trees and forests goes back to my childhood, of course. I was born in the prairies, and the trees we had nearby were mostly in the stream bottoms—hardwoods, willows, maple and types like that—or they were stands of trees that were about four species planted by the homesteaders.

By the time I was a child, these trees were fairly big. So we used to play in them constantly. We used to climb around in trees just like little squirrels. And from that time on, I've always had a love for trees.

Then when we moved out here, the conifer forest became my second home. We were living close to unlogged conifer forest. When I think back, they had all types of different species. But to us, coming from the prairies, the fact that there were so many big conifers stood out. A lot of the time I would spend out in the woods by myself for hours climbing in the trees and sitting in little openings, and we had nest areas, and we'd build huts and things like that.

Gradually you get to know more and more about the incredible beauty of the intact, wild forests, and you develop a love and passion for them. It's an easy, logical step to caring for them to the degree that you become a social activist. So by the time I came to Victoria, which was 1968, I was already a social activist with respect to environmental philosophy and issues.

DT: When I came out to British Columbia, one of the first things I did was begin to take people up to Merve Wilkinson's Wildwood, which eventually became a prerequisite for all the introductory classes in environmental studies. We would all go up to visit the Wildwood and hear Merve Wilkinson. The students were absolutely enchanted with not only Wildwood and the experience of being on a site that has been able to retain the spectrum of forest values over the years, but also by Merve Wilkinson, the stories going back to the late 1930s and 1940s.

Can you set the scene for people who don't know Merve Wilkinson and Wildwood?

DT: Merve Wilkinson has just celebrated his 95th birthday this past September [2008]. Merve has had a piece of property south of Nanaimo on Vancouver Island. It's approximately 137 acres [55 hectares], it's relatively small, but he has been able to maintain the values of the forest not only in terms of biodiversity values, but in terms of it being an uneven aged forest, it has old-growth in it, it has multiple species, multiple ages. He's been able to do that consistently since the late 1930s and have it on a sustainable basis.

In other words, he will only log the amount that grows on a yearly basis. What he's been able to do has received international acclaim. He's had programs on the BBC, CBC, David Suzuki. And it's become a model for the type of ecoforestry that we began to try to disseminate in the 1990s.

AD: He used selective approaches to removing trees.

DT: He calls it selective sustainable forestry.

AD: But he also uses that land for a multitude of other purposes. It isn't just logging that he does.

DT: He had his sheep, for example. He called them "the Brush Cutters." And they would be let out at certain times of year, particularly in the summertime, to cut down the brush because you know that sheep will take everything, even seedlings in the winter. So he'd make sure to keep the sheep penned off in the winter.

He would, for example, not cut the very largest of the Douglas firs, recognizing that the squirrels would go after the best cones on the fir and the squirrels would in turn become the planters. He wouldn't engage in manual planting, but allowed natural seed dispersal to take place with birds, with wind, with squirrels.

He would maintain habitat for flicker populations to help keep down bug populations.

So his approach is multidimensional, recognizing that it's a managed forest, but as a managed forest he tries to learn from nature's wisdom as much as possible, then have it applied.

▲ Can you explain the concept of wild foresting?

DT: We're using *foresting* as a verb rather than as a noun. *Ecoforestry* is a noun. *Wild foresting* is a verb. It relates to a process. It relates to a systemic way of looking at the world without a radical separation between the person and the forest.

You're not looking at a forest just as a series of objects out there. It's no longer a commodity. Instead of an "I, it" relationship, it becomes much more of an "I, thou" relationship. It's a way of being that on one level recognizes that we are inextricably linked with the health of the forest ecosystems and our health is linked with that.

For a very long time, forestry has been engaged in making the ecosystem change its requirements to meet the needs of an ever-expanding economy. Now we have to flip that around 180 degrees and make the economy a subsystem to the imperatives of the ecosystems. It recognizes no separation between living sentient beings and ourselves and the larger ecosystem. So it's a way of being, it's a way of living, it's a way of thinking. It's a recognition that the so-called environmental crisis isn't a crisis so much out there as a crisis in our values, in the way we think, in our mindset. And the mindset we have profoundly affects the landscape, the mental typography, the physical typography, and there is a profound interrelationship. We wanted to get that idea across in the very nature of the term *wild foresting*.

▲ So you've blended the scientific for the spiritual?

AD: Oh yeah.

   DT: Absolutely.

   AD: The first book was art and science. Now wild foresting takes it a step further and brings the spiritual into the equation. We had a subtitle of an article we wrote on environmental ethics, which was, "Seeing tree$ and not forests." And the "trees" had an *S* with a dollar sign.

DT: It was a chapter for a textbook—

AD: On environmental ethics, and they wanted a section on forest values. We were making the distinction between the fragmented, atomistic, reductionist account that just values tree stems in terms of market value, so you just have two-dimensional values here, quantitative values. Whereas a forest has a complex system of values. We were able to determine at least seven major categories of values in a total forest system.

The individual tree can symbolize the entire world. It symbolizes in many spiritual traditions all the dimensions of reality, physical and otherwise. So individual trees, of course, come out of forests. You don't first have individual trees and then forests. Forests and individual trees are part of a dynamic, evolutionary, interrelated process. And in the wild foresting approach, we want to emphasize that even in a city you could have a municipal forest that provides opportunities for wild foresting activities. It might be gathering nuts, it might be going out and sitting in a forest and contemplating.

In Japan, they have this thing called "forest air breathing" [*shinrinyoku*]. They've found out that going out in the forest completely changes the way a person feels about themselves and the world around them. So now they have this concept in Japan as a type of therapy.

DT: This leads into one of the parts of our book, where we're talking about the role of forests in our health. Increasing evidence is now coming out, particularly a lot of work from the University of Illinois, on what's referred to as nature-deficit disorder. Children who are exposed to unstructured playtime in wild areas have much lower levels of AD or ADHD [attention-deficit/hyperactivity disorder].

AD: And other disorders too. Physical disorders—

DT: In studies in Britain, children who go into the woods and engage in climbing trees and what is sometimes referred to as "risky behaviour" have much higher immune systems later in life.

AD: Higher, better functioning.

DT: They are also able to engage in terms of life challenges much better, in terms of having a sense of inner resilience.

AD: One of the original studies on this in North America was done by Edith Cobb: "The Ecology of Imagination in Childhood." She observed a whole bunch of people—exceptional people, they were very creative, they were very resilient and they contributed a lot.

Cobb tried to see what the common background and elements in their childhood were that led them to be so unusual and so creative. The main thing she found throughout their childhood was this similar pattern of a deep connection with natural places where they were allowed freedom to play.

With the new book, are you hoping to reach civilians or politicians or—?

DT: I would say all of the above. The book is done in such a way that it is accessible to the general public, but it can certainly go to policy makers, to educators. We also wanted to have a book that gave a profound level of hope. Each of the essays in that book represents a case study that shows what people at the grassroots are doing in communities all over the world to actually regenerate communities with forest ecosystems.

How do we maintain people's awareness of the fact that the environment matters even when there are shifts in the polls showing environment is not as big an issue for people as the economy?

DT: That suggests that we have been doing a very inadequate job of showing the relation between a healthy economy and a healthy ecology. And the *Wild Foresting* book is trying to get out again that profound interconnection. We're only going to have healthy communities, healthy individuals, healthy economies when you have very healthy ecosystems. There's no separation.

AD: We have chapters on that, on green economics, and on Aboriginal economics and Aboriginal use. All around the world there are probably 5,000 or 6,000 different cultures. The industrial culture that we focus on in terms of global financial exchange is just this tiny bit of the total in terms of the cultural diversity on the planet and so part of what we want to do is create a transition process in our own society to change our industrial culture into a different kind of economy based on totally different energy systems.

What can an individual do? What can I do?

DT: Recognize that at times of crisis and instability, it's also a profound opportunity for hope and transformation. When ecosystems and social systems are highly unstable, that is potentially a very empowering time for individuals and groups.

It's also a recognition that we need to change our priorities. We've been looking for satisfaction, for happiness, in all the wrong places. We need to recognize that.

Recently I was at a conference in Sydney, Australia, dedicated to life satisfaction and happiness, and one of the things that we've known intuitively for some time is that once basic human needs are met, increasing incomes can be a detriment to life satisfaction. Community, wild places, friendships, generosity, appreciation of the aesthetics all contribute to profound life satisfaction. Wild places, wild forests.

If the premier picks up your book tomorrow and says, "These guys know what's going on, I'm putting you two in charge of forest policy," what would you do?

AD: Part of the re-education of our politicians and managers and so on is to stop thinking in terms of megaprojects and large-scale, single-theme kinds of projects and start thinking in terms of the capacity of our individuals to create a wide range of practices at the local level.

With respect to our forests, they have been mismanaged from an ecological standpoint because they have been put under a single model; that is, the plantation model of industrial forestry. So the idea is to cut down all the old trees, which are decadent and so on, and replace them with superior, genetically modified plantations. The goal is to have a continuous stream of material throughput in order to generate a high level of economic activity. But one of the things that people overlook about this process is that the whole thing, like the financial system, is driven by debt. Part of the debt is ecological debt. Part of the debt is local community debt. When I was a boy we had a lot of vibrant local, rural communities in British Columbia, Washington, Alaska and so on. A lot of those are gone now.

The same with agriculture in the prairies. When I was a boy we had all these small agricultural communities, but they were based on complex, mixed farm systems—family farms of 160 acres [65 hectares], 320 acres [130 hectares] or whatever. And they were doing lots of different things on these farms.

In principle, those farms could have been sustained with their cultural complexity indefinitely, but what destroyed them was the financial system. It was the farm policy. There were all kinds of things that were part of this. We're in that situation today where we're seeing distraction of cultural diversity and local communities all over British Columbia because of what's happened in forestry. I can't remember a time when our forestry sector wasn't in crisis. Can you?

DT: No.

AD: It's always in crisis. And it's because of this whole mindset. The large-scale, single-use, single large-throughput approach.

DT: You could say that in British Columbia we need a profound overhaul of the tenure system of this province right now. We also need to be able to allow communities, and the First Nations communities, to have a much greater say on what takes place in the watersheds of their regions.

Also, you have, for example, in parts of Alberta what's referred to as "the genuine wealth indicator model." How does a community determine what it really needs in terms of genuine wealth? And genuine wealth goes far beyond economic wealth. It refers to the wealth of the community, the wild places, friendships, a sense of children being able to play safely, and so on and so forth. When communities can really monitor what it means to have genuine wealth, you see very quickly this leads right into what we're talking about in terms of wild foresting.

AD: It's an alternative to using gross domestic product as the sole measure of economic well-being. It's clear that we need quite different approaches to managing things. We need to include a wide variety of approaches. Not a single-use type of approach. Ecoforestry is a very complex, multivalue, multidisciplinary, multigenerational approach to forest use. And we see wild foresting as another sort of dimension to that, but it includes appreciation for the state of wild forests today and how we can get more wild forests back for people in cities to access.

Oslo has thousands of acres of municipal forests surrounding the city with all kinds of corridors going down through the city to the fjord. And anybody in Oslo has almost immediate access to what they call "free nature" [*friluftsliv*]. Free nature is more like our concept of wilderness or wild nature. But I can have wild nature in my backyard—I do have wild nature in my backyard. I have little creatures come in there all the time that are wild. And I can appreciate them. And the more of this sort of thing we have, the better.

▲ How do you feel about trees?

DT: I love trees. They are friends. They are teachers. I have profound respect for trees.

AD: My feelings about trees are incredibly complex because they are sort of nurtured and formed by my Norse shamanic background. They are part of the central mythology of the north for the conception of the world reality. As you know, this same concept of the tree with its roots and it branches and everything representing all the different levels of reality from the physical to spiritual and the afterlife too is found in cultures the world over. Trees symbolize classification systems and the branching of classifications.

Then on another level there's my deep, personal feeling for trees, and the way that I have felt a special bond with individual trees and also with various stands of forest that I've travelled through in my journeys as a climber and also as a journeyer in the old Norse tradition. The phrase *a Viking* was originally a verb, which meant journeying spiritually, shamanically, in the world, out on the ocean, in the forest. Wide journeying and trees are always part of my heritage. Making things out of trees—building cabins, building forests, making tree houses. My whole life has been influenced by trees. And today and every day I try to go out walking in the forests around here, and a lot of times I'm all by myself—

DT: And with your dog.

AD: My dog of course is always looking for things. And bonding with those trees and receiving a certain kind of communication from the trees. It's just endless in complexity and quality and beauty.

The Ecoforestry Institute Society at www.ecoforestry.ca
The Trumpeter: Journal of Ecosophy at trumpeter.athabascau.ca

15
ANTONY MARCIL
FOREST STEWARD

One of Harry Potter's greatest magic tricks was letting the world know about the Forest Stewardship Council (FSC). When J.K. Rowling announced that the final installment of her beyond-best-selling saga was going to be released on FSC-certified paper, it was hard to miss the existence of an international organization founded in Toronto in 1993 to look for better ways to manage the world's forests. Since then, the FSC brand has become the label consumers and manufacturers look for to see whether their wood is environmentally kosher.

Antony Marcil took over FSC Canada as president and CEO in 2005. Before that, he spent 15 years as president and CEO of the World Environment Center—an organization that pioneered working with industry on environmental issues. "The dialogue at that time was not *with* industry, it was *at* industry. The WEC was founded to break that mould and basically sit down and say, 'Look, if industry are causing the problem they've got to be part of the solution, and the only way they can be part of the solution is to sit down at the table and work with them.'"

In 1997, Marcil was included in the first worldwide listing of "The Top 100 Figures in Environment, Sustainable Development and Social Issues" by *The Earth Times*. He served a one-year stint as planner in residence at the School of Planning, Faculty of Environmental Sciences, University of Waterloo, and devoted five years to an unexpected way to save the planet: tax reform.

Among his triumphs with FSC Canada: convincing the Government of Ontario to commit to switching to FSC-certified paper.

I met Marcil at the FSC office in downtown Toronto in September 2007 to talk about the history and future of the FSC, how tax reform could save the world and why he'd trash Canada's Ministry of the Environment.

What is the Forest Stewardship Council?

At a really basic level, it's a collaboration among Aboriginal peoples, environmentalists, labour, church groups and the private sector to bring about the proper management of forests. It had its roots in the failure of the Rio Conference in 1992, when a draft forest treaty was proposed and the governments of the world did not come to any agreement on how to protect tropical rainforests. So there was a meeting held just a few blocks from where we're sitting here in Toronto in 1993 saying, well, if the governments of the world aren't going to protect forests, let us do it. So it's a real grassroots, democratic movement.

What are the FSC's biggest triumphs?

Surviving. In that first room when people came together, there were people whose goal was to stop all logging, just as there were industrial loggers there whose goal was to keep on logging. To get opinions as divergent as that to actually come together and agree on standards and principals to govern logging is an incredible accomplishment. I think the mainstream logging industry expected the whole thing to sink under its own weight. And probably it should have, but it didn't. The result is that we've got standards that are respected worldwide and are in force. There are more than 80 million hectares—that's 200 million acres—of forest land around the world certified to FSC standards by independent auditors—just like independent financial auditors, there are independent forest auditors. [By May 2009, that figure had risen by one-third.]

So the triumph is to have gone from 1993 to today and still be in business and have a growing bank of certified forests. More important is to have marketplace recognition and that's there now. All the seven major banks in Canada use FSC paper at one level or another, and more companies are coming on board all the time. Companies know what it's about.

How is a forest certified? How is a company certified?

The starting point is to have a set of standards. The FSC does not write the standards, FSC manages the process by which stakeholders write the standards. Our job is to make sure that all the stakeholders are represented around the table. So the first step is to have Aboriginal people, environmentalists, labour, other civil society groups and industry sit around a table and write the standards.

We have up on the wall the 10 principles that govern every standard around the world. The task of that working group around the table, all those disparate interests, is to keep looking at those 10 principles and ensure that they respect each principle as they write the standards for how forestry should be done in one of our four forest types. Once those standards are written, they're submitted to the accreditation arm of FSC in Bonn, where technical people ensure that they do properly respect the 10 principles. There are 57 criteria to satisfy each of those principles, and then there are indicators in terms of how you measure whether the criteria are being satisfied. That's what's behind the FSC. If we didn't have those standards, we wouldn't be in business.

Once the standards are in place, our job is to get forestry people who own or manage forests, whether they're small or large, to voluntarily agree to adopt the FSC standards as their rules for cutting. So we have four basic forest types in Canada: British Columbia; the boreal—which goes from coast to coast;

the Great Lakes–St. Lawrence—which is not quite finished; and the Maritimes.

We have a couple of hundred forest management certificates in place. Tembec, for example, was the first major forestry company to undertake voluntarily in 2001 to have all its forests audited. Similar to hiring a financial auditor, they asked for quotes from auditors accredited by the FSC—there are five auditors who work in Canada. From the quotes, they chose an auditor and the audit took place. The job of the auditors was to go in and actually verify on the ground, in the field and in the records in the offices to see whether Tembec's performance measured up to the requirements of, say, the Boreal standard.

Once that was done and the certificate was issued, Tembec then has trees that it is logging each year that are FSC-certified and can have a logo on them. Those trees will either go to a sawmill or to a pulp mill or to some other use. And we have what we call a chain of custody . . .

When an FSC label shows up on a ream of paper that you or I buy in Staples, how do we know that the fibre that's in the paper that's in that ream actually came from a well-managed forest? The answer is that the logo on that package has a number on it, and that number is a certification number of whoever made that ream of paper. You can call them up and say, "Can you show us the documentation as to where you got the fibre that you used to make the paper that's in this thing?" And they would then have to say, "We bought it from so-and-so pulp mill and their certification number is . . ." And you could actually trace it all the way back to the forest.

▲ **What is the history of the other groups that have come up to certify forests? Was the FSC the original one? Are the other groups more industry based?**
The FSC was the launch pad, if you like. The initial reaction of big

industry was bemusement, expecting this group of rabble-rousers to sink under its own weight, as it might well have done. A year later it hadn't sunk, and it was still moving forward with defining the principles, and a couple of major trade organizations, one based in Canada and one based in the States—I have no idea what their actual motivation was—each created standards. The American Forest & Paper Association (AF&PA) actually hired people within their own staff and they sat down and wrote standards that did not require a third-party certification auditing. The Forest Products Association of Canada (FPAC) was a bit more subtle. They hired the services of the Canadian Standards Association, a very reputable organization that has been in business for a long, long time writing safety standards around electricity and plumbing and all kinds of things, to write a forestry standard.

How big of a deal was it for the FSC when J.K. Rowling announced that the final Harry Potter book was going to come out on FSC paper?
It was a big deal for the very crass reason, if you like, that it was a big quantity of paper on the one hand, and at the same time it was the kind of thing that might garner some media attention because, as you well know, good news doesn't play well in the media. And the FSC and the whole notion of certification has not had a lot of media coverage; therefore, an announcement of that nature is very good.

That struck me as a phenomenal PR hit for the FSC.
In a broad sense, in a public sense, it was. In Canada we've had in place a strategy since I joined in March 2005 to target business to business partly because certainly on the paper side there wasn't enough product out on the marketplace to garner a general PR campaign, such as the Potter announcement to the general media. There was very little that you or I could do in response

to that announcement. You'd read the story and say, "Oh that's wonderful, I'm going to do my bit," and you wouldn't be able to find anything. Whereas business could find anything because wholesale, on the business level, there's lots of product available. So that's what we've been doing, creating buzz from business to business, feeling that would give more time for more product to be available at the retail level. And today there is. Lowe's and Rona and Home Depot, Ikea—you can find a fair number of products with FSC labels. And you can go to Staples or the Printing House and buy reams of paper that now have the FSC label.

▲ **So this is now something that a regular consumer could do. I could go out and say, "Hey, I'm going to make a choice. I'm going to build my house with FSC wood. I'm going to print this story on FSC paper"?**
Yes. It's uneven though; in some parts of the country you would have a hard time finding products. Everything you need from dimension lumber to MDF and plywood and other forms of panel material and so on, all of it exists, it's just not easily available in all parts of the country yet. But that's changing rather rapidly.

▲ **Where does Canada stand globally in terms of honouring forest standards?**
In terms of the FSC, we're the leaders. We have the largest amount of certified forests. We have 22 million hectares. The next leading country behind us is Sweden with 12 million. We expect to have 30 million hectares certified by the middle of next year. From there it will be a matter of convincing companies that chose one of the industry standards to switch to the FSC.

▲ **Going to high school in BC, we would always talk about the forests as a renewable resource. I wonder with what we know now, what we're seeing now, with the value of old-growth**

versus the value of tree farms, is forestry really a renewable resource the way that I would have understood it as a kid?

You touch on a point that is actually a sore point with us, which is the liberal use of the word *sustainable* by the industry standards. Both the American-based and Canadian-based industry standards—the Sustainable Forest Initiative in the US and the National Sustainable Forestry Management in Canada—call themselves sustainable. We don't use the word *sustainable* unless it slips in by accident. If you look at our materials you won't see it because we won't know. Our kids may know whether forests that were FSC-certified were in fact sustainable in the sense that the ecosystems have survived intact. There are a lot of people who believe you can't do any logging in a forest and have the ecosystems survive intact. In 50 to 100 years from now, we'll know whether FSC standards resulted in sustainable forests in that intact sense. But I have no doubt that in certification that takes place under many systems—because there are many systems around the world—it's inappropriate to use the word *sustainable*. You'll see for example a lot of advertising being done by logging companies in the States that are certified to a different standard than ours, saying to the general public that forestry is in good shape because every year the acreage of forests in the US is growing. Well, sure you can have more tree canopy when you look down from an airplane, but that doesn't mean that there's anything alive under the canopy other than trees and to me, that's not a forest. So it's not an issue of what percentage of the country is covered with trees, it's certainly not the number of trees—there are some companies that advertise, "We planted two million trees last year." So what? The fact that one planted two million trees says nothing about whether the result of that is a healthy forest.

If somebody came to you tomorrow and said, "Hi, we're putting you in charge of Canada's forests," what would you do?

Run the other way? It depends. In the classic way of thinking, to be the minister of forestry you're not in charge of Canada's forests. If they came along and said, "We're going to make you minister of finance," I'd be there in a nanosecond because then I would be in charge of Canada's forests. The minister of environment, minister of natural resources, minister of forests, they're not in charge of Canada's forests. They're in charge of exploiting them. They're not in charge of maintaining them.

What could you do as minister of environment?

Nothing. I think ministries of environment should be abolished. They're a crutch. They're something that allows the general public to feel that, "Oh yeah, we've handed that responsibility over to somebody. They're in charge so I don't have to worry about it now," which is a total crock. The people who can influence the environment are the minister of industry, minister of trade, minister of agriculture, minister of public works. Those are the guys who have an impact on the environment, not the minister of environment.

So what would you like to see done?

Abolish the ministries of environment. Make the responsibility for the condition of the environment a line responsibility in all the ministries. And put the minister of finance in charge.

How do you feel about trees?

We couldn't live without them. You look back at all the societies that have become extinct, you'll find for the most part it's because they changed the condition of the land that wouldn't support growth of vegetation anymore and therefore they lost their water supply.

Forest Stewardship Council at www.fsccanada.org

16

JOHN VAILLANT
GOLDEN SPRUCER

When I first read John Vaillant's story about the Golden Spruce in *The New Yorker* I had two simultaneous thoughts: what an incredible story and who the heck is John Vaillant?

I don't know the work of every writer in Vancouver, but I couldn't believe there was someone in my city who wrote this wonderfully whom I'd never heard of. The article is one of the best I've ever read. It's an amazing story, beautifully told. But the book that article launched rocked my world.

The Golden Spruce: The True Story of Myth, Madness and Greed shares the story of a 300-year-old Sitka spruce that was a scientific miracle and a spiritual one—an albino tree with golden needles that never should have existed, never mind survived. And in 1997, a logger cut it down. What made the book an instant Canadian classic is that Vaillant wove the adventures of the tree and the man who cut it with the story of BC's forests and the way trees have been chopped for generations.

Vaillant grew up in New England, and he and his wife, Nora, first visited BC in 1993, connecting with the cultural history at the Museum of Anthropology in Vancouver and the Royal British Columbia Museum in Victoria. They moved to BC in 1998 when Nora was accepted into graduate school at the University of British Columbia in Vancouver.

When I set out to talk trees in the summer of 2007, Vaillant was the first person I contacted. Most interviews with him focus on the man who killed the Golden Spruce, but I wanted to know what a writer from the US saw in the death of a tree in Haida Gwaii that every writer from BC had missed. We talked at his kitchen table in his home on the west side of Vancouver.

How did a travel story on kayaking turn into a book about a mythic tree?

Completely accidentally. I'd been called to the islands just by their location, hanging there off the coast and looking so mysterious and inviting. But I really didn't know what was up there, beyond historic totem poles. I had this out-of-date tourist map that had various sites to see and things to look at, and one of these was a golden spruce, and I thought, "That sounds really bizarre." When I was done my kayaking story, I had time to kill basically and rented a car on Graham Island and was driving around looking for this tree. And you know that maze of logging roads up there, which kind of spiders out like varicose veins across the beautiful island? I was lost in those and found somebody who looked like they knew where they were more so than I did, and I asked them, "Where is this tree?" And he said, "You can't see that anymore."

"What do you mean?"

"It's been cut down."

And I knew it was protected. I knew it was special in some way. I said, "Who would do a thing like that?"

He said, "Well, a logger did it. But he did it as an act of protest."

"Against what?"

"Against logging."

So I'm scratching my head. "What happened to him?"

"Well he disappeared in a kayak, and nobody knows where he is."

And I just filed that away. There was a story. It wasn't the story I was there to write. I was completely blindsided by it. It had been three years since the tree was cut down, when I first showed up there in 2000. So this guy had digested it. He wasn't angry anymore, it wasn't news to him anymore. His matter-of-factness and the strangeness of it, that combination really just lodged in my mind. When I got home, I just started scratching away at archived newspaper articles from that event, and everything I learned just raised more questions—it basically got stranger and stranger.

▲ **What was it that caught you most? Was it the story of the Spruce or was it the story of the logger?**
The combination of things—of this extraordinary tree that really shouldn't have existed, that was almost a botanical impossibility—

▲ **Can you explain that?**
Chlorophyll is what makes trees green. And chlorophyll is what makes photosynthesis possible. And photosynthesis is what makes it possible for all the green things on Earth to live. This tree had lost most of its chlorophyll. It had been burned out. It had a genetic defect that caused the chlorophyll to break down under ultraviolet rays. So basically you had this gigantic tree that needs chlorophyll to grow, but there is no chlorophyll there, except underneath it's got maybe 50 percent of a normal tree's allotment of chlorophyll.

So it would be like you or me trying to compete against successful, huge, powerful athletes—say in a marathon, or a football game—with one lung, or half a heart, or something like that. Imagine being debilitated to that degree, and then transpose that onto a tree, and you would think this tree would be a dwarf, it would be some stunted version of itself or it would die very early because nature is quite ruthless with genetic defects. And instead, this tree thrived.

It was as big as the other trees. It was an appropriate size for its age, 300 years. And it was absolutely healthy appearing, even though it was growing strangely. It was deformed in terms of its shape as a Sitka spruce. It had these bizarre pigments. And it was apparently sterile—it produced cones, but none of the seeds were viable in the cones.

So you've got this tree with these three massive defects—massive in terms of what makes the natural world go round—and yet there it was standing there, bigger than life. Then you have on top of that this spiritual narrative significance to the Haida, this social environmental significance to the anglo logging population of that place—for whom it was kind of a mascot. Then you have the scientific puzzle behind it. And then you have this guy.

One of the things that captured my attention when I saw this in *The New Yorker* was I was aware of the Golden Spruce being cut when it happened and still have the newspaper clippings in my files of amazing BC stories and yet no one had taken this on. I thought it was fascinating that here you were—you weren't from BC, but you saw this and thought, "Yeah, this is special." What do you think you saw that everyone else missed?

I had an advantage being a stranger to this place in that I was absolutely astounded by the enormity of the trees, by the intensity of logging, by the intensity and sheer enormity of the landscape. Locals are used to that, they've adjusted to the size. I'm coming from a tiny little state—you can drive across all of New England in a day—and my first trip through BC took 16 hours and I thought, "Wow, I've seen *some* of the province." And I had just got to Lillooet.

When I went back and looked at the map, I had to take a deep breath. My whole psychic molecular concept of what nature and space is was completely disrupted by moving here. It's a whole different scale. I was already kind of in awe of this place, and then here was this bizarre tree and this strange man. And the other thing that I had going for me was that I wasn't on any sort of deadline. Newspaper journalists have a terribly difficult job; they need it yesterday, or at the very latest tomorrow. It took me six months of freelance scraping and digging and calling and pestering just to find out where Grant Hadwin had grown up. It took six months to find that out! Who has that kind of time? Also, there was something about him that resonated for me. For a lot of people, he was so close, just to think about him made people angry. He took something beautiful away. He did a terrible thing. And why do you want to go and honour a guy like that with curiosity? I think that really came through with the newspaper articles. But again, being from outside, I didn't have the same stake, I didn't feel violated the same way that local people felt. In some ways I was more open, and probably more familiar with psychological complexity than maybe the average journalist is.

What were some of the bigger surprises for you looking into the history of forestry in BC?

I was coming from a kind of lefty, enviro, urban background. I was coming from a very simplistic point of view: trees are good and logging is bad. And I know that was simplistic. I was aware enough of myself to realize that my point of view seriously limits my understanding of these issues. So I really wanted to get into the other mindset—the mindset of the Haida, the mindset of a forest, insofar as it has a mindset, the mindset of the logging industry—and see where they were coming from. Also, coming from a pastoral place like New England, it's all been cut down. You call them clear-cuts here, and back there we call them fields, but they're the same thing—it's a clear-cut that's been planted with something or other. Clear-cuts and wide open spaces are normal to me. When I first got out here I felt—and I still feel—often totally oppressed by the forest. My impulse is: I need some light! How am I going to get that light? I'm going to cut some space for myself. I feel that deep inside and that is the way Europeans think.

And all the people who came here and logged this place, they all came from the east. All of them came from places that had been clear-cut hundreds, if not many hundreds, of years prior. So they were recreating their own world. Also I was really fascinated by this sense that these forests were worthless and lifeless and obviously they're anything but, and we're just beginning to understand that now really in the late 20th, early 21st century.

But up until this point, they were described as environmental deserts, ecological deserts. The wood was described as useless. They were just these enormous 100,000-ton obstacles to progress, and they had to be removed. And I knew those were my origins too, so I had to challenge them, explore them and try to understand them and I think readers can relate to that.

Looking at the history of logging in this province, what do you think we've learned from it?

I don't know whether we've learned anything. It seems to be business as usual. A few pockets have been saved at enormous cost. I think human beings have incredible difficulty with balance. There is also this attitude that you really see here, that nature is like a sucker to be taken advantage of and you can just go in and rob that cookie jar any time you want.

This notion of seeing nature as a business partner, I think that's our only way out—as a partner in our future. Right now a lot of people have trouble seeing it that way. There is still that legacy of contempt: "This is mine, I can take this. It belongs to us, we bought it, we leased it, that's what it's here for." I think underneath that, there is a kind of contempt and disrespect and I don't think it's even conscious, but it's an old, old attitude, hundreds and hundreds of years old, and it's laid waste to enormous tracts of this planet.

▲ **I heard the other day about an environmental philosopher saying we need to rewrite the social contracts to say it's no longer just man-and-man, but man-man-and-nature.**

There are a lot of people who are stressing that in different ways. I was trying to think, "How do you articulate this to an Exxon or a Monsanto or somebody like that?" And that's why I came up with the idea of a business partner, because if you screw your business partner, that is going to have repercussions for you in the future. Either you're going to be closed out of markets, or you're going to get a bad reputation or whatever. If you show respect to this place that gives us so much it's just astounding, it's the *sina quo non* of everything. Yet we have this ability to objectify it in this way that, even if it's not intentionally so, has hostile effects.

▲ **Can you tell me about the *National Geographic* story you're off to do? That sounds like another amazing BC forest story that isn't known.**

The Muskwa-Kechika is an enormous segment of the northern Rockies, bracketed around Fort Ward and Fort Nelson. It's 25,000 square miles [65,000 square kilometres], about the size of Scotland or the state of Maine. It's reputed to have 50 pristine watersheds in it—all the species that lived there 300 or 400 years ago, to my knowledge, are still there and thriving. A couple of fellows named Wayne Sawchuk and George Smith and some other concerned citizens up there realized that in order to maintain the integrity of this place, they needed to build a consensus of behaviour and a set of guidelines for how to act in there. And they engaged people from the oil and gas industry, people from the timber industry, big-game hunters, First Nations, environmentalists, local land-use planners, and said, "Okay, can we agree that this is a place that we need to keep intact?" And people said, "Yeah."

And how are you going to get oil and gas out of it without destroying it? How are you going to hunt grizzly bears in it? It was too big just to close down and say, "No one's allowed in here." Wayne Sawchuk himself is a trapper, he's a former logger, and he had an awakening. Grant Hadwin had an awakening too and it took a much more destructive turn. This guy has seen the writing on the wall—it's all over BC—and has said, "Okay, that's going to happen here, unless we act proactively and engage all the stakeholders." It's been in existence for 10 years, there has been some exploration but it's been very, very sensitively done, and it's such an enormous area that it presents a model in microcosm for how to manage ourselves on a provincial scale, on a national scale, potentially even on a global scale.

And the fact that these guys have been able to keep this conversation going without anybody blowing up or disregarding them and just kind of storming in there and logging a couple of alleys just because they can—that hasn't happened, to my knowledge. So what's strange is that nobody knows about this place.

I've asked many British Columbians who are environmentally literate, who read the newspaper, who are well travelled, and they don't know where this place is. I'd never heard of it until I met Wayne Sawchuk, who was one of the founding fathers of the idea. And when he told me what he was up to, I was just amazed and I said the world needs to know about this vision and this alternate way of engaging stakeholders that isn't acrimonious, that isn't defensive, that isn't fearful.

▲ **If somebody turned around tomorrow and gave you a magic wand or a magic piece of golden spruce, what would you do with our forests?**
I would be overwhelmed by that responsibility. I would call a moratorium to cutting old-growth, right off the bat. I would put

a lot more time, money and resources into understanding the systems that function underneath and within our forests. I think they're just seen as fibre farms and it's much more complex. We're seeing it if you look at the bug kill in the interior, we're messing around with stuff and we're reacting to things that we don't understand terribly well.

I think we all owe it to ourselves to spend a lot more time trying to understand it. But I don't have a quick, magical, clean answer. One of the things I learned, from the little I know about this, is that it's enormously complex and a lot of it is dependent on region and species and terrain and all kinds of things. But one of the things I'd want to do is not export raw logs.

If you look at a fir beam or a Sitka spruce beam—now the only ones who can afford things like that are millionaires. They're luxury items now, they're status symbols—beautiful, clear pieces of wood. I think that's healthy in a way. I think we should treat these as precious, beautiful and extraordinarily valuable objects. And the forests that create them should be revered and respected accordingly. We finally got to that point, but partly it's because there's so relatively little of it left. You can look at reports from 1919 and people were saying, "We are logging this stuff faster than we can replant it, and we're not going to see this stuff again for 800 years, if ever." People have known this for a long time. It's so tied to our appetites and our sense of what we're entitled to.

I almost want to take the logging industry and do this massive workshop where people explore their sense of what they have a right to, what they think they should have, how they see the forest facilitating their dreams and their lifestyles. You know, how many acres of forest is a speed boat worth or a jet ski or a mortgage or a college education? And is that the most efficient way to use a forest? These are questions I don't really see anybody asking. So, I don't have a clean answer . . .

But the other thing is, you'd be surprised if you speak to a logger or someone in that industry who's not been pushed into a defensive position how deep the understanding is of what they do—how well they understand the cost of their actions. And yet, they're caught in it. We all are. The way we're tied to oil, they're tied to logs. None of us want to give up our car. And a logger would be an idiot to give up making 600 or 800 bucks a day to finance his dreams, whatever they may be.

How do you feel about trees?
Pretty awe-inspired, really. I just look at those things, and they're able to make so much from so little. All these invisible things—air and sun and tiny minerals and water—and to manifest such gravitas and such generosity and such beauty out of that. What a model for how to be in the world.

John Vaillant at www.myspace.com/johnvaillant

17
JOHN WIGGERS
TREE GURU

Before filming *The Green Chain,* I decided that the movie should be connected with a group that was building bridges between loggers, environmentalists and First Nations. I knew the group existed, but I couldn't figure out what it was called. I sent a dozen emails to friends and strangers describing this group and no one knew what I was talking about. Then I got on a plane from Vancouver to Toronto and sat next to this guy who started pulling pictures of trees out of his briefcase.

He told me he had been in Vancouver for a meeting with the Forest Stewardship Council of Canada. I opened my laptop, popped open one of the emails I'd been sending out and asked him to read it. "I think I've been looking for you."

And I had.

John Wiggers and I spent almost the entire flight talking about the movie—which Wiggers seemed even more excited about than I was. We also discussed the FSC, turning green and his pilgrimage to Haida Gwaii to see the remains of the Golden Spruce.

Wiggers is a woodworker, a craftsman who creates beautiful specialty—often one-of-a kind—pieces from his workshop in Port Perry, Ontario, and he's devoted his life to working with ethically sourced exotic wood. He felt that seeing the Golden Spruce was important.

Sharing the story was more important.

He'd promised a Haida elder that he would spread the story of the Spruce. A few years later, in May 2009, I interviewed Wiggers for the book and he informed me that the flight he was on was the flight back from his very last board meeting with the FSC—which he'd just left after spending four years in a variety of positions including treasurer and chair—and he was, and still is, looking for new ways to help raise awareness of the issues facing our forests.

For *The Green Chain* movie credits, I declared him "tree guru." And the title delighted him.

What fascinated me about Wiggers wasn't just that he loved trees, but that trees clearly love him. On three occasions his life has been saved by trees—once, he and his brother were saved from drowning by a tree branch; on another occasion his car miraculously "threaded the needle" between a pair of trees to avert a wreck; and on a third, a bolt of lightning hit a tree right next to him. That last incident —and Wiggers' visit to the Spruce—were the inspiration for the executive character in *The Green Chain*.

Here's Wiggers' story about his journey to see the Golden Spruce and how his four-year-old son turned him green.

▲ I know you had a huge turning point in your life when your son asked you about the type of wood that you use. Can you talk about that?

My son was four and he was watching a cartoon on TV [*Captain Planet*], which was about all these environmental superheroes going around the world saving the planet from various disasters. In one episode, a furniture maker had a machine that was eating up the rainforests and a conveyer belt that was spitting out all kinds of finished furniture at the back end of it. To save the day, all these superheroes came in and beat up the bad-guy furniture maker, blew up the machine and saved the rainforest.

My son knew I made furniture for a living. So he started asking me questions like, "Are you one of the bad guys, are you destroying the forest?" To try to answer honestly to him, I began to realize that I didn't know with 100 percent certainty exactly where all my wood was coming from.

Of course that was late 1980s, early 1990s, around the time

when the Brazilian rainforest was in the news a lot. And there were some fledgling initiatives out there to try to come up with answers and find solutions. In those early stages, this ultimately is what prompted me to get more involved and take a closer look into what I was doing and what kinds of wood I was using.

▲ **Is that how you got involved with the Forest Stewardship Council?**
The FSC didn't form until 1993. Early on, it was grassroots groups like the Woodworkers Alliance for Rainforest Protection, which goes by the acronym of WARP. They had some early efforts. There were guys like Andrew Poynter and Scott Landis, who would take trips into Brazil to look out for certified operations that were doing a good job and arrange to purchase these materials. But it was very hand to mouth, like a drop in the bucket. Then these guys started trying to figure out how to manage the logistics, and one of the things that came up fairly early was that there was no set of standards in place to really determine what was good forestry in Brazil compared to, say, British Columbia. Obviously these are two very different forest types and what constitutes good forest management is very different in various forests around the world. Soon it led to the recognition that there was the need for standards development, which led to the creation of the FSC in 1993.

▲ **How did your visit to Haida Gwaii to see the Golden Spruce stump come about?**
In early 1997, the tree was cut and it ended up making headlines around the world. I happened to catch it in the paper and there was just something about reading it that absolutely rocked me. It was like a kick in the stomach, and I made a promise to myself that if I ever got out to the west coast of Canada I would make a visit to go see the tree—I guess you could say to pay my respects.

I never really gave much thought to getting out there, when I would, how this whole thing would come together. You know, one day it will happen, one day I'll go.

On some level that decision set off a whole chain of weird coincidences because it was shortly after that that I read about Smart Wood and the FSC program. I got involved with that, which put me in touch with the FSC people in Toronto, which in turn led to my being elected onto the board a couple of years later. As it turned out, the first board meeting we had was in Vancouver. So all of a sudden I had the opportunity to go to the west coast, and I just remembered what I promised myself three years before. I started putting together a side trip to go there after the board meeting.

A lot of roadblocks came up. There were two First Nations people on the board who gave me a few phone numbers and email addresses of some people on Haida Gwaii who could help me. But the emails did not work and the phone calls were not being returned. It just looked like that part of the trip wasn't coming together. But then, a few weeks before my trip, a buddy of mine came by and asked how things were going.

I told him, "I'll be going to Vancouver for the meetings, but I won't be seeing the Spruce because I just can't seem to pull anything together." So he said, "Well, you're not going to believe this, but I was at a restaurant last night with friends talking about your trip, and somebody at the next table overhears me and hands me this business card and said, 'If your friend's having trouble getting out there, have him give me a call.'" It turned out the person was the marine stewardship coordinator for Haida Gwaii and was living in Massett, not far from where the Golden Spruce was felled—so all of a sudden I had a contact person setting me up with a place to stay and transportation. Of course, my problem then was getting a plane. I called a travel agent—and found I couldn't find a flight there on such short notice. Apparently there's one little plane flying in and out of Haida Gwaii every day and everything was booked solid. Once again it looked like a wrench in the works, then half an hour later my phone rings and it's the travel agent saying, "There's been a sudden cancellation on the flight you were looking for, so if you want it, it's yours." So I said, "Book it."

As I said earlier, it was like an unseen hand at work. Things just magically pulled together, and I was able to make that trip.

▲ **Can you describe the experience of seeing the stump?**
First, you're driving down this pretty long road. The forests on either side are actually just little strips of forest, you can get glimpses through little openings that a tremendous number of trees were being felled. Finally, you get to a spot in the middle of nowhere with a trail leading in. There was a sign that said, "Golden Spruce" with an arrow pointing up the trail, but someone had painted out the word *Spruce* and put in *Stump*.

I went down the trail and thought I was actually going to end up walking right up to where the tree was. But it wasn't until I got to the edge of the Yakoun River that I realized the Spruce was on the other side. So that was a pretty disappointing revelation to get there and find out I wasn't even going to get up to the actual tree. I actually just stood for a minute by the edge of the river and started laughing my ass off. I thought, "This is so ridiculous." I went through all this effort to make a trip to go out to see this tree, and now I'm separated by 100 yards [90 metres] of water and not even close to the tree I came to see. It was kind of surreal.

But standing there, just laughing about the whole absurdity of the situation, it kind of brought the question to mind of why did I come here at all? What was the whole point? I mean, I could see the fallen tree on the other side. It was more than three years dead at this point. It was just brown and dead, hanging over the edge of the river. There was certainly no opportunity for me to go there. So I just stood there in a silence and really thought, "What was the point of coming here? Why all the effort to see a dead tree?"

And it was in the moment that I looked back up the trail that brought me in. And to get to where I was standing I actually had to climb over a tree that had only fallen in the last day or two, probably because of a storm. It had fallen right across the trail, so I had to climb up over the trunk to get to the other side. Looking at the tree, it looked about the same age as the Golden Spruce would have been. It was like the two of them had grown up together, but on opposite sides of the same river.

So I made my way back, and just stood there and took in the silence for a while. I got a little lost in the whole nothingness of the place. Then I figured that this was as good a place as any to pay my respects, so I did. At one point I was just looking down and there was a pool of water where the tree had been standing, and looking down into the water I saw these images reflected onto the surface. It may sound like a pretty spacey thing to say, but there were four very distinct images reflected on the surface of the water. I even ended up taking a photograph—and some parts of the images came through—but it was mostly distorted with the flash I had. The images were quite apparent, very detailed and beautiful in some way. But what was the point? What did it mean? Who would ever believe me if I told them? I figured if I said anything people would think the cheese had slid off my cracker.

So I just kept it to myself for about a year. I went back home; back to work; and said nothing. It wasn't until later the next year, it was after 9/11, the first anniversary of my trip was coming up, and the memory was always gnawing on me. It was getting to a point where I felt like I had to capture the memory. I decided to have it painted to canvas to hang in my office as a reminder of the experience. It just had that kind of a powerful effect on me. I ended up commissioning an artist I'd met at a craft show the previous summer. She caught me off guard, insisting not only on doing the painting, but not being paid for it. At that point I felt selfish about keeping it for myself. So we decided that instead of making it for me, it would be given to the Haida as a gesture of hope over the loss of the Spruce.

The following spring at the Forest Leadership Forum in Atlanta—this is now April 2002—the FSC brought out an elder from Haida Gwaii, who was Leo Gagnon—to see the painting on behalf of the Haida. It was at this point I spoke publicly for the first time about my experience there. Something about the story must have moved Leo very deeply, because by the time I was done, he was in tears.

Are you comfortable talking about the response when you handed over the painting?
Let's just say there was some mixed reaction, which is to be expected. But like I said, Leo was deeply moved, so it seemed obvious that the gesture was a positive one. Leo talked about what was happening on Haida Gwaii, the kind of devastation taking place in the forests. He didn't say it in his speech, but afterwards he and others came up to me saying that the falling of the Spruce was to many indigenous peoples akin to 9/11; one guy said to me, it was the falling of their tower.

For Leo, he made a connection to an ancient Haida prophecy that said if the tree is allowed to fall, then the world as we know it will soon follow.

You said also that the images you saw were significant.
After the presentation, the painting was packed up and sent to Haida Gwaii, and once it arrived in Masset there were other elders who came out to take a look at it. One of them commented that

it looked like a spirit was being pulled from the fish. In my mind I was seeing separate images. There was a separate head connected to the salmon. By the one interpretation, the way one elder was viewing it, he said it looked as if the spirit was being pulled from the fish. Then, a few weeks later, the salmon failed to return to the Yakoun for the spawn, which really was a shocking revelation from what I understand because salmon are integral to Haida life and culture. So salmon not returning to the Yakoun was a very upsetting event. Oddly enough much of the problem had to do with falling trees and silt clogging the river.

▲ **If somebody put you in charge of our forests tomorrow, what would you want to do?**

Hire someone better qualified. Like I said before, I learned a lot of things while on the FSC board. One of them was there are many experts who know a lot more about the forest and how it's properly managed than I do.

▲ **How do you feel about trees?**

I love trees, to steal a catch phrase from your movie. My affinity with trees on a personal level is very deep. On three occasions, a tree has saved my life. So obviously there's a connection that has a very personal meaning for me.

John Wiggers at www.wiggersfurniture.com

18
CHIEF BILL WILLIAMS AND NANCY BLECK
WITNESSES

For 10 years—from 1997 to 2007—people were brought into Squamish territory to visit the Elaho Valley in the hope that they'd fall in love with the pristine forest and join the groups determined to save the area from logging. At least, that was the original intent of photographer Nancy Bleck and mountain climber John Clarke. But when the pair met Squamish Chief Bill Williams, their mission became much richer. Instead of simply experiencing the beauty of nature, the 12,000 visitors drawn to the Elaho over those 10 years shared the culture and hospitality of the Squamish people and became "witnesses" to help raise awareness that this valley was not just a potential park or a future tree farm, it was Squamish land. (That concept wasn't just a challenge to the government, but to the environmentalists as well. The Western Canada Wilderness Committee had named the land the Randy Stoltmann Wilderness Area in honour of the then recently deceased BC environmentalist and author.)

Bleck, Clarke and several other photographers documented these life-changing weekend visits, and their awe-inspiring and awe-capturing images were displayed at the Roundhouse Museum in downtown Vancouver.

When we spoke, Bleck was compiling a book to document the history and process of the Witness Project, "to share it with the public as a model for what happened, because it's a very interesting community collaboration."

Chief Bill Williams is serving as one of the First Nations hosts of the 2010 Olympics. "We invited the world to come to Vancouver in

2010 to be a witness of the Olympics. As each country comes in, 130-odd countries, we will be doing a drumming and singing ceremony and welcoming them to our traditional territory."

I spoke with Chief Bill Williams and Nancy Bleck in April 2009 on a bench outside the Roundhouse Community Arts and Recreation Centre, where we talked about the Witness Project and the importance of being called as a witness.

Can you please introduce yourselves?

BW: I'm Telálsemkin Siyám, Squamish Nation. Telálsemkin is my ancestral name, Siyám is a designation given to me by my family and recognized by the community. The Indian Act calls me "a chief." I am 1 of 16 hereditary chiefs in the Squamish Nation.

NB: My name is Nancy Bleck. My adopted Squamish name is Slanay Sp'ak'wus, they call it a *ninahalahan* name, which means it not an ancestral name, it's like a nickname, a name that was given to me here at the Roundhouse in 2001.

What does it mean?

NB: It means "woman eagle," so Eagle Woman. Slanay is woman and Sp'ak'wus is eagle. I'm one of three cofounders of the Uts'am–Witness Project that ran out of the Roundhouse from 1997 to 2007. There was also John Clarke—his name is Xwexwsélkn, which means Mountain Goat—and Telálsemkin Siyám, Chief Bill Williams. The three of us proposed the Uts'am–Witness Project to the Roundhouse and they accepted it as their first residency. It was originally going to go for about six months.

The year before Witness began, I met with Chief Bill Williams on the sandbar of Sims Creek, which is about three hours north of here. I was working together with John Clarke for the year before that. John and I had met in 1995, and we started thinking about how we could get more people to recognize what was happening to the land in the Elaho Valley.

We were at a camp out there through the Western Canada Wilderness Committee. After about 100 attendees went home, there were about five of us left. We were sitting around a campfire just brainstorming about what we could do. Our idea was to bring people up to the land to see for themselves what was happening with the clear-cut logging. We thought that there was no better way to get people to care about what's happening than to show them first-hand. We didn't feel that the political manoeuvre of the last stand, that kind of plea to the public to say, "let's protect this forest," would work anymore in that particular climate. We thought that was giving in to a "war in the woods" mentality, and we wanted to steer away from that. We wanted to go more in the direction of providing a direct experience with the land, as a way of getting more people to have a relationship to it. So our strategy was to strengthen relations to the land, and who better than John Clarke himself, the legendary mountaineer, who had 600 first ascents to his name. So we became very passionate about bringing people up to the Elaho Valley.

One weekend we were gathered on the Sims Creek sandbar to thank all the people who had volunteered for us that year, and Chief Bill Williams was there. I had noticed him in the crowd and realized that nobody knew who he was. He had actually offered me his cell phone in 1996 when I was looking for a reporter, and back then nobody had cell phones, it was really high-tech stuff. I was desperately looking for a reporter at the Squamish Petro-Can, and this person just handed me his phone and said, "Here, you can use mine."

So I approached him on the sandbar and asked him who he was. He gave me his business card, which said, "Hereditary Chief of the Squamish Nation." I was very surprised and at the same time immediately recognized that we had never asked permission to be on this sandbar in the Elaho Valley.

We were bringing people out to the land, to Sims Creek, Tree Farm Licence (TFL) 38, as the forestry industry may call it. But we had never approached the Squamish Nation, even though it's their traditional territory. I asked Bill whether he approved of what we were doing and he said he liked what we were doing. Then I asked if he would address our group and welcome us, and he said, "Well. I like what you're doing, but I can't formally welcome you here because you've all used the logging road to get here, and that in itself is a breach of protocol."

So that was my first lesson in protocols.

But he did greet our group and he was very open and he did like what we were doing, and I practised saying his ancestral name for at least 15 minutes and it was really hard to practise Squamish. So when I introduced all the volunteers and I gave them their gifts, I thought it would be appropriate and important to present a gift to one of the chiefs of the Squamish First Nation. And I happened to have—no kidding—an eagle feather with me that weekend, and I presented him with an eagle feather and addressed him by his Squamish name, Telálsemkin Siyám, and I think that was the first time that we started seeing that there was a way we could work together.

▲ **Why was the logging road a breach of protocol?**
BW: Because nobody had come to Squamish and asked to be a part of the land. Nobody had come to talk to us and let us know what was happening. Nobody had asked for any kind of ceremony to be done.

When anything on the land is done in terms of development—building a road, building a house—there's always ceremony to thank the ancestors, to thank Mother Earth, to thank the trees and let the trees know that they are going to be taken down and that some good things are going to happen to the remains of the trees. There was no ceremony at all. And I couldn't in all consciousness go up and say "Welcome to our traditional territory" when every protocol you could think of to get at that spot had been broken.

▲ **Who built the logging roads and why hadn't they approached you—wouldn't that be a standard thing to do?**
BW: It is a standard thing to do today. Right up to that point, it was standard to ask permission from the provincial government and the provincial government would give the rights and then the people would just move in and do whatever they thought they were doing with no protocol toward the First Nations at all.

▲ **When did that shift? And how did that shift?**
BW: It shifted in 2001, 2004, 2007, with the Gitksan-Wet'Suwet'en court case, Delgamuukw, with the Haida court case and then with the Taku. They kept on building on each one, and as time went on, the Supreme Court defined more specifically what the provincial government was not doing with regards to protocol in recognition of Aboriginal rights entitlement for us as indigenous peoples of Canada.

▲ **Had you been actively campaigning against that particular TFL?**
BW: Not in any sense of real petitioning. We were asking the TFL to include us Squamish people in terms of the development to create jobs and opportunities for our people. From May to September in the back pages of *The Georgia Straight,* there was an invitation to all people who wanted to come up to the wilderness area in Sims Creek. And we were negotiating with the province of BC and they were saying, "This is TFL 38, this is our rights there." And then I hear it from WCWC that, "This is Randy Stoltmann land and we want to create a park over here."

People were layering all kinds of labels on the land there. So I was going up there on that long weekend in September and said, "No, no, no. If you really, really want to know the name of this land, this is Kwa Kwayexwelh-Aynexws, this is what we named this place 10,000 years ago."

What does that mean?
BW: It means "transformation."

What's the history of that name?
BW: Do you have four days? It's a beautiful, long complicated story, like a lot of our stories. But that's why I went up there, to say, "You guys are naming this place Randy Stoltmann. You guys are naming it TFL 38. You guys are naming it 'the Sims Creek environmental whatever.' It's not all that. Let's get down to the basics and define the Aboriginal rights and title to the land. Let's take a look at the real substance of the land and the real connection to Mother Earth."

Was this a new battle for you, or had you fought this over other pieces of land?
BW: It was a battle that my grandfather fought and my dad fought and I fought and all us hereditary chiefs are always trying to define and let people know that we are not extinct and we're not going to hide away in a museum. We do have our language intact. We do have children being raised on the land. And we do know what is happening on the land.

NB: It's a complicated layer. What you're saying is by naming we claim. So in 1997, we had TFL 38, we had the Randy Stoltmann Wilderness Area and we were the group with no name that brought people out there to see what was happening, but it had always been the traditional territory of the Squamish Nation. And that was the side that people weren't seeing.

▲ **I'm intrigued that it never crossed the minds of the environmental groups.**

NB: That, for me, was where I departed from the environmental community in the sense that it had set itself up in that kind of dualistic thinking of the "war in the woods" and the "us versus them" strategy, the "loggers versus the environmentalists." That was a strategy I didn't want to get into. I had just come back from Prague, where I had lived for three years, and I saw the process of change happen there through the Velvet Revolution. What I saw and witnessed there was people like artists and workers, the intellectual community, coming together and creating that change. And I thought, "Why is it that we label so much when it comes to the environment and BC?" We put these labels—you have to be an environmentalist or a logger or a First Nations person or something.

So I thought, "Where is everybody else? Where is the intellectual community? Where is the artistic community? Where is the scientific community? Where is the township of Squamish?" And why do we have to have these conversations that are always so rhetorical, argumentative? We had death threats the first weekend we went to Sims Creek. John Clarke was in hiding in the back of the truck. As we were going through one of the logger's blockades, one of the loggers said, "Have you seen this man? Have you seen this white-haired guy?" Our driver said, "No." And the logger said, "Well, death to him." It was very, very, very violent in those years, and people did get beat up up there. So the actual opening up of a ceremony out on the land was a way to engage in a very complicated conversation that went across languages and across ideology and across history.

▲ **How many images were included in the Witness project?**

BW: Thousands. One of the things that we found very interesting was that we had no idea it was going to last; we didn't know that it was going to last 10 years. We were asked by the Roundhouse to have a registration list, people had to sign up. So we know by that sign-up list that about 12,000 people went up there and some were part of the Witness project. Some people found it so wonderful to go up there that they went up five weekends in a row. It was quite exciting and different. One of the things that was very different was the direction of Uts'am. Uts'am in our language means "to witness." And to be a witness in our ceremony is very sacred. When you're called as a witness, the people in the community that you are in recognize that you have a sound mind, that you are intelligent and that you will be called as a witness because of that. And once you stand up and are recognized in our community, you are to look at what is going on, why the speaker has called your name and you are to listen. That is your job—just to watch and to listen. It is not to bang nails into a tree, it is not to go and drain gas from a logging truck, it's not to do anything but sit and watch and listen. That's what a witness is.

And that's what I had to try to explain to the people that went up there. In order to be invited into our house—because in every ceremony that we have, we invite people to our house to be a part of that event—they're part of our family. And to be part of my family, I respect you as an individual, and in return I want that same respect back. I am a very strong witness in our community. So if I call you as a witness, then I expect you to act accordingly. Some people who went up with us did not want to act accordingly. But the members of the group around us talked about it in a long dialogue—I mean, from like nine o'clock at night to about four in the morning—of what a witness is. I would go to bed and sleep, and they would come back up and say, "Chief, Chief, can you explain this?" at two in the morning.

To be part of Uts'am–Witness is showing a part of who we are as a group because we have a very oral history and that's how we move our names down from generation to generation, that's how we move tracts of land from one family person to another. It is not by some notary public's document. It is done in a ceremony.

▲ **I asked Nancy the meaning of her name. What does yours mean?**

BW: My name was given to me by my grandmother and it was her great-great-great-great uncle on her father's side. So last time Telálsemkin was walking around was probably in the 1600s.

▲ **Do you feel Witness saved that area?**

BW: No. What Witness did was the job it was supposed to do. It made people aware of whose land they were on. It made people aware that there is ceremony to the land. The transformation began in 2000 and was finalized in 2006 when the provincial government recognized the Squamish Nation's land-use plan, which signified the ancestral names of all parts of our traditional territory. And when they did that, they also put a map out with our ancestral names on it. Unlike in 1996, by 2006 the provincial government knew where Squamish Nation traditional territory is loud and clear. And all the bureaucrats know that if there's any development on the land within the Squamish territory, then they have to talk to the Squamish Nation.

▲ **Can you see Witness being applied anywhere else?**

NB: I think there are a lot of places in the world that struggle with indigenous rights to land. I think they also struggle with the coming together of diverse communities with different interests in the land who speak different languages and have different histories. That notion of coming together through ceremony and through having a common goal is really at the heart of it. It was not an easy dialogue in the beginning. As we mentioned earlier, there were death threats. We managed to bring loggers into the conversation. We managed to bring people who had no idea what it means to be part of a project that involves the environment, First Nations history, arts, science and the political economy of forest industry. That's not an easy conversation to get into. So I think other parts of the world can benefit from our asking, "How do you sit down and how do you have those conversations?"

▲ **What's your land-use plan and how was that devised?**
BW: We went to the community and told them that we have to reach out to the provincial government and reach out to all our outside communities to let them know the real title to the land, the real names of our rivers, the real names of the mountains and identify generally where our sacred places are in our traditional land without talking about which ceremonies are done in those sacred place. We needed to let people know that these places are sacred and that in order for you as a non-Squamish person to understand what is sacred about it, you have to come and live among the Squamish and participate in our rituals to determine that.

▲ **If somebody said that tomorrow "you're in charge of all the forests in BC," what would you like to see happen?**
BW: I would tell people that the five-year plan that the province demands to have from each of the logging companies is a bunch of crap, that we should have a 200- or a 500-year plan for our forest. In order for us to be able to utilize the forest properly, it has to be old-growth forest, not an 80-year-old forest that the provincial government recognizes as a second-growth cut standard.

▲ **And how do you feel about trees?**
BW: They're good. They're valuable. They're very useful for the

home and the community. And we can use trees, but we also have to be able to do ceremony around the trees to let the forests know that they're doing a good job, let the forests know that in order for us to do things like cut down trees, we need the forests to protect the trees and understand that we will continue to do ceremony well into the future.

NB: There's one thing that I remember from when I interviewed Bill a few years ago. He said, "Well, now that we've won the fight of saving the forest, we have to work on the bigger fight of introducing that to our youth."

▲ **How's that fight going?**

BW: A good one. Because like any youth, they have to understand, they have to be informed, they have to realize that it is part of our culture, it is a major part of our culture. And for me to explain, again it's another two-hour story about what the significance of a tree is.

The Witness Project at www.utsam-witness.ca

19
WADE FISHER
UNION MAN

When I set out to start the podcast series, I asked Wendy and Tom McPhee to name the best person to talk to about trees. They said I had to meet Wade Fisher.

Fisher had just been lured out of retirement to represent forest workers on the Cariboo-Chilcotin Beetle Action Coalition (CCBAC) and to chair its forest worker strategy sector. He'd previously participated in the Cariboo-Chilcotin land-use planning process and co-chaired the Cariboo-Chilcotin Economic Action Forum. By the time I met Fisher in Williams Lake in the summer of 2007, he had withdrawn from the group. He explained that while this was the right issue for him, it was the wrong time for him to take on a new cause.

Fisher started working in the bush near Williams Lake in 1969, cutting trees and setting chokers [a choker is a short length of steel cable wrapped around the log to help haul it off-site]. He left the woods for the plywood plant. He also drove skidders and did just about every mill job there is, including running edgers, barkers, chippers and loaders. Then he left the mill for the union hall and the union hall for meeting rooms filled with politicians and planners.

In 1979, Fisher became an officer in the union local for the International Woodworkers of America. And he spent the next 30 years as one of labour's most respected voices in BC. For five years he worked as a negotiator for the "brown" (industry) side, representing more than a dozen groups in devising the Cariboo Chilcotin Land Use Plan. "We were the only group in BC at the time that had a negotiated settlement to the war in the woods. We did that with the so-called greens. I don't like those labels because I think they're pretty unfair to both sides. I know some people who call themselves environmentalists and would be quite insulted that you considered them 'brown' because they happen to believe in some form of industry on the land base. So I think the labels are wrong, but that was a very interesting process."

Fisher helped the browns and the greens create "a land-use plan here that I think could have stood the test of time and would have provided long-term sustainability for a whole number of groups out there—right from tourism operators to the back-country guys, the more upscale and the forestry guys. I think we could have had it all." But Fisher says the plan was another casualty of the bark beetles.

"The problem I see right now with the bug is we're overharvesting. We know we're overharvesting, and I'm not sure that's a wrong decision, but I know what it does do is that we're going to have a falldown in forestry that nobody thought was possible. We're going to overharvest so much that it's going to take a long time to have timber that's usable. We're going to have logged way more than we should have."

Fisher and I met in Wendy and Tom's living room to talk about vanishing trees, vanishing lifestyles, Dogwood Diplomas and how the provincial government's policies could have an even more devastating impact on workers than the beetle infestation.

What's it like to be a union guy working in BC's forest industry these days?
In the last few years it's been a pressure cooker because there's nothing good happening. Every situation you're in, even if you do something that's good, it's good only in that if you hadn't done something you'd have lost 15 jobs. So you've done something and you get to say, "Well, boys, we've saved five jobs." But you've still got 10 people going down the road. So it's been very tough in the forest industry.

How does the union deal with that?

The union wanted to have an element where you could look at pensioning people off to keep jobs for younger people. They wanted to do some pension bridging. They wanted to have some money for training in advance, so you don't wait until a guy's on the street to start training him. We ran an education program here a few years back and we know from that that if you train people while they're actually working, it's cheap. But if you wait until a guy's unemployed and you've got to support his family while he's training, that's real expensive. If you're working and I can pay for your courses, and you can do your courses while you work, that's pretty cheap.

We know that a lot of our people need a lot of education before they're ready to take any kind of meaningful training. You take a guy who's worked in the forest industry for 25, 30 years—he's 48 or 49 years old. He probably got out of school at 17 years old, if he stayed that long. Many guys in that age bracket didn't stay. They were out when they were 16 years old and working. So to take that guy and say, "You're done here today. You're going to go take some training"—well, he doesn't even have the basics to take the training.

So a few years ago, under the Harcourt [NDP] government, we started a course, which ran for three years. We got funding for it and we were actually the first to do it. Then they started having them all over the province. But in our case the basis of our program was that you train people while they're working and you don't differentiate between what kind of training they can take. Our approach was that the people who need training the most aren't the people that start taking training. We wanted to create an atmosphere in the workplace where the highest paid and most skilled guys were taking training, so the other guys looked at them and said, "If that guy's doing it, it must be good." So for the higher skilled guys, who had their grade 12 or maybe a university education, we wanted them involved in training so it brought others along.

We offered Dogwood Diplomas. We even put a guy through as a commercial pilot, which we took a lot of flak for in Victoria. They said, "Why'd you do that?" Well, he was a highly skilled guy, and we wanted those guys involved in training, so we wanted him in. He wanted to become a pilot. He had a pilot's licence and wanted to be a commercial pilot. So we arranged to have him get his instrument rating. Once he had his instrument rating, he was away to the races, because the guy put a lot of effort in himself. So we've done a number of things like that.

What was really amazing about that is we had a person in this community who couldn't read or write—couldn't read or write—who went and got his Dogwood Diploma in three years and graduated. That's the kind of thing that we and the union movement wanted to see.

Tell me about that Dogwood Diploma.

The Dogwood Diploma is exactly the same diploma you'd get if you go from grade 1 to grade 12 in the school system. It's not like writing a GED or the equivalent. You actually end up with a diploma from school district 27 that says you've completed grade 12 education. That's a Dogwood Diploma.

So the guy's got a diploma that's good anywhere in the world. It's recognized. We even hired teachers from the school district to help tutor our guys. And it was all done on an individual basis. You could sign up, and you could work. We had one guy who just wanted his grade 12. He got his Dogwood Diploma and went on to do advanced mathematics.

Is the pine beetle the scariest thing out there right now for workers?

I don't believe so.

What is?

The scariest thing is the fact that the government has given up on forest workers, in my opinion. Government has given away the farm to the big companies.

How so?

When they removed appurtenancy a few years ago—and people will tell you that appurtenancy never saved a job—

Can you please explain the term for me?

Appurtenancy was a section in the Forest Act. It was a social contract that forest companies signed when they got tenure in this province. What that social contract said is: you'll get timber, but for that timber you're going to provide jobs in communities to help build rural communities and make them sustainable and all of that good stuff.

But this government we've got today under Gordon Campbell came along and removed that appurtenancy section. Their argument—and you couldn't stand up and call them liars because what they said was true—was that appurtenancy by itself never saved one job. Appurtenancy alone did not. However, when it was in place, it helped prevent situations where if you owned tenure and said, "I'm shutting this mill down, and I'm going to truck these logs to the next community," the government could step in under the Forest Act and say, "Hold on, you've got to prove to us that you're not economical. If you can prove you're not economical, you can transfer your timber. But you've got to prove that."

The government would come in with a specialist in the industry, look at the books, look at what the company was doing, and in lots of cases what happened in this province is they'd come back and say, "The company can run, that company can be profitable, and here's the recommendations to fix what they're doing."

We had mills, in Revelstoke for instance, that would have been shut down 10 years ago—I believe it's down today—but it wasn't shut down 10 years ago because the government came in with specialists, looked at what was needed and offered the suggestions to fix it. The guy who owned the mill could have said, "No, I'm not doing anything," but if he'd done that, he'd have lost his tenure.

When the government took that away, what they basically said was, "You've got tenure, you've got it forever and the price you were supposed to pay to the people of British Columbia, you no longer have to pay." It's like me going out and getting a mortgage and committing to pay the bank, and all of a sudden the government comes along one day and says, "Oh bank, you're not getting that money from that guy. He gets to keep his house, and he doesn't have to pay you." That was the price we, the people, were supposed to get for tenure. We're not getting that price.

▲ **What was the official reason for doing this? And what do you think the reason was for doing this?**
Well, the official reason at the time—it's been a while now—was that they were streamlining it, making it less red tape, and they said that it didn't work anyway. But under the previous government, the Harcourt government, there were several mills—the Squamish mill, I think, was another that was dealt with under appurtenancy and kept running at the time, and it was just shut down here this year. Or last year, I guess now.

But the problem with this is it allows for companies to not create jobs in communities. It allows them to move wood wherever they choose to move it. It allows for the sale of that wood. And they say, "Well, that would happen anyway." But the reason it allows for the sale of the wood for export—I'm talking about export here—is you create no demand for the wood. If you have no demand for the wood, you can apply for an export permit.

If you allow people to shut mills down that are economical, it doesn't take very long before there's no demand in the area for the timber. Once there's no demand in the area for the timber, you get to apply for an export licence.

So this is about exporting raw logs to the States?

You bet it is. I've been away from it for a while now and I don't remember the exact figures, but we had enough wood exported out of the Terrace area to run a mill, the size of one of these big mills here in town, for more than a year. In one fell swoop, 300,000 cubic metres just gone for export because there was no demand in the local area for that wood.

So instead of sending it from Terrace to, say, Quesnel, Williams Lake or Prince George, they were allowed to ship to the US?

You bet. It's a matter of record.

Now why wouldn't mills turn around and say, "Forget about Williams Lake, forget about Quesnel, we're going to build one gigantic mill in Kamloops"?

That's a great example because that's already happening in Prince George, it's happening in Quesnel. West Fraser just built a mega-mill in Quesnel. Canfor built a mega-mill in Houston. And those mega-mills just haul wood in from miles and miles and miles. What they don't tell you is that local communities are dying because of this. Canfor just shut a mill down in Mackenzie—permanently gone. But that's happening all over the place in this province where communities are going to die because the mills are shutting down, and they're getting to move logs wherever they want with the government now having absolutely no say in how that's done.

It used to be the government had a say—you had to have the right logs, the right mill. But they were able to control the movement of logs so you couldn't just export the jobs out of a community. Today there's no barrier. So consequently, what you're seeing in the forest industry is an example of this.

The big are getting bigger. The small are disappearing. They're virtually gone. Just the odd one. We've got one success story here in Williams Lake—the Sigurdson brothers, they're doing well, they're an independent. But the independents are disappearing out of this industry. So consequently, you're seeing a fairly major change and we're just seeing the tip of the iceberg.

If somebody came at you tomorrow and said, "Wade, we're making you minister of forests, you can do whatever you want," what would you do to fix forestry in BC?
I would certainly try to bring back appurtenancy, or some form of control. Appurtenancy was a word that people didn't know or understand, that was part of the problem, I think. But I think you need to be able to use the levers of power in government to make sure that communities survive and that business survives. They both have to survive. But you can't have one survive at the expense of others. That's what's happening in rural BC today—business is surviving but the communities and the people in them are going to be in real trouble.

We're just seeing the tip of the iceberg with this mill closure in Mackenzie. I think we're going to see a lot more of that.

How do you feel about trees?
I think I love trees, actually. I used to tell Dave Needs [negotiator for the green side]—and to me it's kind of tongue-in-cheek actually—that the difference between him and me was that he loves his trees vertical and I love mine horizontal. When you work with people you've got to poke each other now and then. And that was sort of our jousting session . . .

But from a very personal perspective, trees have been my livelihood. My father worked in the forest industry. So all my life they've been our way of earning a living. They're pretty near and dear to me. And I think that's why this bug thing is so devastating, because if you've been involved and know what's going on out in the land base, you know we're overharvesting, we know there's no replacement for the trees we're taking now, because we're taking them too soon and we're taking too many. You know that you're seeing the end of a way of life for a lot of people. So that's the sad thing about it.

Trees have been my lifeline. They've clothed us, fed us, put my kids through school. You know, my son's in the forest industry and he's a young man with a family and I look at the situation and say, "What's he going to do?" What replaces that for those people? And there's tons of them out there. Look around at all the people in the forest industry—and what have we got to offer them once this forest industry goes down the tube? And I think it's going to go down the tube in a major way.

There will always be some forestry. It won't totally disappear. But it's going to become such a skeleton of its former self that you won't recognize it.

Canadian International Woodworkers Association, now part of United Steel Workers Canada at www.usw.ca

20

TOBIAS LAWRENCE
SOCIAL CONTRACTOR

In the 2009 BC provincial election, Tobias Lawrence ran for the New Democratic Party in Prince George–Mackenzie against the province's minister of forests, Pat Bell. And Lawrence ran on a platform made out of trees. For Lawrence, forestry and the way BC manages forests is a lifelong passion.

She started her working life as a treeplanter, then worked in a sawmill, where her jobs included bin attendant, stacker operator and Level 3 first-aid attendant. She also worked for the BC Ministry of Forests in timber sales.

But her life changed when she volunteered to join the campaign for one of BC's most colourful and controversial members of the Legislative Assembly, Corky Evans—a former cabinet minister whose passion was forestry. Lawrence clearly made an impression with the NDP because that led to a stint as the assistant to the NDP's forestry critic and, later, a party candidate.

The more time the left-wing Lawrence spent working in and around forestry, the more interested she became in a concept invented by BC's proudly right-wing Social Credit Party, a concept with a seriously left-wing name: social contract. For Lawrence, the BC Liberal Party's biggest sin in forest management was ripping up that contract.

A few days before the 2009 provincial election, Lawrence was appointed the BC chair of the 100-year-old Canadian Institute of Forestry—a group dedicated to raising awareness of the importance of forestry in BC. I interviewed her that night at her Prince George home about her love for our forests, the people who work in them and the death of the social contract.

You've called forestry "the backbone of the province." Is it still? Or is that back broken?

The back is broken right now. And the people of British Columbia have to realize the value that forestry has to the province. If we don't have a strong forest industry, then what do we have as a province? It used to be the huge economic generator that provided us with our health care facilities, educational institutions, roads and infrastructure. If we don't have that, where does the revenue come from for the people of BC?

What can we do? What does the future hold for us?

I think we need to go back to a thing called "social contract." Social contract and the legislative foundation were radically changed in 2003—along with the removal of appurtenancy [from the Forest Act]. That change has contributed to 57 mill closures and 30,000 job losses in BC. Our government went from a direct economy to a market economy, and it was partly in response to globalization and partly in response to the economic philosophy of the Liberal government. They did it hand in hand with the forest sector because the sector was demanding it. The sector made a bargain with the government that if you release the constraints, we will bring you economic prosperity because we are so good at this. The result of this—well, our forest industry is now called a sunset industry. All you have to do is drive around this province and in every community you will see closed mills. Communities like Mackenzie started to shake in 2006, long before the economic downturn.

Can you explain social contract?

Tying logs and jobs to communities so that communities share in the revenue from the trees. I wrote a paper about this in 2005 when I was studying to get my degree in forestry.

If a mill is in your area, that community has likely been built because of that mill, which employs local people. So people spend money in their own community, which keeps the community viable.

Before 2003, if you wanted to shut down a mill, companies had to sit down and negotiate with the government. The government and the minister of forests would say, "Okay, if you want to shut down we're going to take back some of your timber licence, and we want you to reinvest in another community for job creation."

▲ **So is social contract the same as appurtenancy?**
Sort of. Appurtenancy is a part of the social contract. So is tenure, which was designed in the 1940s to protect forest-dependent communities from the boom-and-bust cycles of the forest sector by introducing sustained-yield forestry and encouraging investment by large companies. With tenure, companies paid "stumpage fees" for every tree they cut and a lot of that money was invested in the areas where the trees were cut.

Social contract promised communities that they would share in the value and benefit of logging in their area. So appurtenancy is about keeping mills in specific towns, but social contract refers to taking care of both the towns and the regions surrounding them.

You have to remember that about 94 percent of the province's forests are publicly owned. The BC government manages the land in the public interest, trying to balance environmental, economic and social issues. That means the province is the major owner and the major supplier of wood to the forest industry. So logging generates revenue for government. On the other hand, our elected government is responsible for developing law and policy dealing with the forests of BC. And that's why the social contract is so vital. In a community like Mackenzie, there are no mills running right now and the social contract is broken.

▲ **When people in BC talk about dying logging communities, the first one everybody points to now is Mackenzie. You were**

campaigning to represent the people of Mackenzie. What was that like?

Mackenzie was heartbreaking. The first time I went there was in November to door knock, and it felt like a Stephen King novel. Houses were either for sale or empty.

This was once a vibrant town, a town that accounted for 5 percent of BC's gross domestic product. A town that once had the highest household income per capita in BC. A town where now you can rent a house if you can afford to pay the utility bills.

There used to be six mills in Mackenzie. Mackenzie's forest district is the size of Vancouver Island, and Mackenzie has a forest that is green, not devastated by the pine beetle. And these mills are allowed to close?

What happened here?

Women would stop on the street and ask me to bring their husbands home because their men left for Fort McMurray or Chetwynd to find work. Not wanting to leave and hoping to see Mackenzie return, the women and children stayed.

There was one gas station in Mackenzie, and one day there was no gas. Greyhound will be shutting down. It's hard now to recruit doctors, nurses and even teachers.

We went to eat in a restaurant during the winter and we could not take our jackets off, and I wanted another jacket. They did not have their heat on before opening. I poured a coffee not to drink but to wrap my hands around it, to keep them warm.

The minister of forests, Pat Bell, made a statement at a rally in May 2008 where 1,200 people showed up to fight to save Mackenzie and to shed light on their town and to show that they did not want to go quietly. Bell said that he would help get flights direct from Mackenzie to Fort McMurray. That was the brilliant solution to save Mackenzie, a direct flight out of our province. I wonder what W.A.C. Bennett would say about that.

I also heard stories of families breaking up, divorces and even suicides. It was awful. I met the son who found his dad hanging from a rope in his basement. This was the reality of Mackenzie.

If somebody put you in charge of BC's forests, what would you want to do?

Take care of our forests and bring the forests of BC back to the people of BC—bring back research, bring back silviculture. Help BC become a province where we are planting trees again, doing research, brushing, spacing and pruning, tending our forests. And not only that, but moving toward not looking at a tree as a piece of a two-by-four but looking at it as an ecosystem, and moving toward ecosystem management of our forest land base.

Your new job is to raise awareness of BC's forests. What else do you think the world should know?

They are important. Trees contribute to our climate, our environment, carbon sequestering. The dead trees right now are emitting a lot of carbon, which is not good for climate change. It's really important to get it out there that a tree is not just something that gets milled; it's something that contributes to your community, your clean air and your environment.

We've got to get planting trees because planting trees is sustaining us for the future and the generations to come, for our children and grandchildren to have something in this province.

How do you feel about trees?

Forestry is my first love. When you break everything down, it all comes back to our forests. Our forests were once the backbone of this province. Our forests built this province. If you care about our forests, and know what that really means, you care about this province. I love trees.

Canadian Institute of Forestry at www.cif-ifc.org

21
AVRIM LAZAR
COMPANY MAN

As soon as I heard that the president and CEO of the Forest Products Association of Canada—the group that represents between 60 and 70 percent of Canada's major forest companies—received a standing ovation at a recent Green Party convention and plans to save Canada's forest industry by making it the most eco-friendly in the world, I had to talk to Avrim Lazar.

Lazar spent 25 years in the Canadian senior public service. "I was a policy leader. I did policy on things like child poverty, sustainable agriculture, I did environment policy, I led Canada's development of a position for Kyoto. I did labour and market policy. So I spent 25 years trying to figure out what sort of policies would help Canada, and then I decided to go to the other side of the table and represent the private sector."

Lazar and I spoke in May 2009 about switching from public service to the private sector, his passion for the environment and his master plan to save Canada's forest industry by turning it green.

▲ **How did you end up at the Forest Products Association?**
I had a long interest in both economics and the environment. The forest industry seemed to be ready to find a way of integrating these two elements, so when they approached me, I said I'd only be interested if they wanted to make environmental progress a major theme in how the industry transforms. And they hired me. So here we are.

▲ **Why was environmental progress so important to you? When did you become environmentally passionate?**

I grew up canoe tripping and hiking, and my idea of heaven is hanging out in the woods. And I still do. This afternoon I'm going to pack up for a weekend canoe trip with my two sons. And mine and my wife's idea of a holiday is getting way out in the wilderness for weeks at a time. I've always had a spiritual connection to nature, a real sense of how much it nourishes my soul. It may seem surprising, but if you talk to most foresters, they actually went into the profession because they love hanging out in the woods. So even though their job is industrial, they chose it because of where it allows them to live.

Can you talk about your action plan to save the forest industry?
The forest industry has to go through several transformations. One is to reorganize its structure to be more efficient. But another is to anticipate the world's market demand and what will be sufficiently in scarce supply to allow the BC forest industry to prosper and for us to keep jobs in rural communities.

One of the very key things that we can see coming is a demand for products that have the highest possible environmental credentials. So the idea is pretty simple—make as rapid progress as we can on our environmental performance and then market that progress as part of a market differentiation plan.

In the world, there is huge deforestation and huge problems with illegal logging. In Canada, we haven't been deforesting, we've learned how to live within the forests and we've got no illegal logging. The industry has embraced the need to do something about climate change and taken on an ambition to be carbon neutral without purchasing offsets. So we think we can actually turn this into something that keeps jobs in rural Canada.

There's another way of looking at it, which we're beginning to find intriguing. If you look at any of the current trend lines, they don't take humanity anywhere we want to go, whether you're looking at toxins, climate or what's happening to biodiversity. So the world is going to be looking for new models for the production of material goods. In the forest world, there are basically two models at play. One is the plantation model, where you clear-cut a large piece of territory and basically create a tree farm. And you can do it in a manner that's environmentally responsible within the concepts of industrial forestry, just like you can have industrial production of organic products. These tree farms have bio-engineered trees, they're basically clones—row after row of clones—big mono-cultures force fed chemicals, and they can be very efficient.

Another way of thinking about forestry is to say, "Can you practise it in living, natural forests?" In Canada, we approach this not out of genius or anything, it's just that that's how we've always done it. You live in a rural community, you cut down some trees, you sell them and you make a living. And over the years we've learned to do it more and more responsibly, more and more respecting ecosystem values, more and more being careful of erosion, more and more sensitive to endangered species and biodiversity. But I'm beginning to think it's not how responsible we are that in the end is going to be the model for the world—others are doing that too. I'm beginning to think in the end it's the concept of living within nature, the concept of practising forestry without destroying the forests to do it, without destroying all the natural biodiversity and planting a tree farm.

The best estimates are that global gross domestic product is going to double in the next 20 years. If you project out any of the trend lines from industrial models to that kind of doubling, it's not sustainable. There's no way of imagining it's sustainable.

I think humanity isn't going to want to go back to living in caves. People are going to want well-being, and certainly those in emerging economies are going to want to see their standard of living go up. And the question of how we can produce the things people want—books, bookshelves, wood to build houses—without further undermining nature's processes, that's going to be one of the critical questions for the future. The practice of forestry in natural forests, instead of industrial forestry through efficient production lines, will turn out to be the answer.

▲ **What received the most resistance in your organization when you started talking about this? I can't imagine that everybody was thrilled to bits when this became policy.**

Well, there's a lag between the action and the market response. So a company takes on the extra expense of improving their forest practices; for example, to be a member of my association you must have your forestry practices certified to the Canadian Standards Association or to FSC or to SFI [Sustainable Forestry Initiative]—all of this requires fairly expensive changes. And the market response to it is always delayed. The same with the climate change objectives we've made. It takes a while until the marketplace turns it around.

But one of the phenomena that has been most interesting as we, and the association as a community, start doing these things is each time you take a step in the right direction, you change your identity a little. You get a better sense of yourself as an individual and as a group as environmental leaders. So psychologically it's a virtuous cycle of doing the right thing, sometimes strictly for commercial reasons, sometimes anticipating regulation. It's not always enlightenment. But you do the right thing and you begin to see yourself as environmentally progressive, and if the community that's doing this talks about it that way, and if they're acknowledged that way, then the next time an opportunity comes it's not as much of a stretch because that's who you are already.

I don't want to make it sound like it's a transformation, you do one good thing and then all of a sudden guys running forest businesses can switch places with Tzeporah [Berman]. It's a gradual process, but it's amazing how it accelerates as each step makes the next one easier. There's that and then there's the simple question of, Jeez, maybe the world will be ready for this and we'll be able to make it into a distinctive market niche. The two play together. Right now the market niche part of it is barely playing out, whereas the "this is who we are" is turning out to be surprisingly powerful.

▲ **Wasn't Canada's brand fairly ugly for a while on the environmental front?**
It was. Partly because, like the rest of the world, we were going about our business without environmental sensitivities. Nobody was that concerned with deforestation or climate change or biodiversity loss 20 years ago. So we certainly weren't ahead of the pack.

There were worse, we weren't like the clear-cutters of the rainforest. But we were nowhere near where we are today. It's partly that and partly we have a very competent, committed community of environmental groups who held us to account. I can't say we enjoyed it at the time, but places where the environmental movement was not as mature or active weren't held up to account in the same way. So that helped wake us up. Let me be clear now, we're still not perfect by any stretch, but we're pretty convinced that we're now as good as any and better than most and getting better all the time.

▲ **Was Clayoquot Sound a big turning point?**
It certainly was one of the more public and dramatic turning points. But when you go there because you're forced, it has less transformative value than when you make changes because you think, "Jeez, maybe I can do this before someone forces me."

In my experience, one of the big turning points was just the commitment in certification. And the biggest one was when pretty well every industrial sector—and this was early on in the Kyoto process—well, every industrial sector was not only finally acknowledging that climate change existed, but that it was the biggest threat to our country and to our industries. And we were saying Kyoto isn't strong enough, it isn't aggressive enough and we're going to exceed it. That for us was a huge transformation point. We now exceed it seven times.

Can you talk about the response when you spoke to the Green Party?

I went there to explain where we're coming from. I'd say the initial reception as I started to speak was polite but not enthusiastic. But as I explained what we were trying to do, we actually ended up receiving a standing ovation, which is actually more than I expected. Not only was I the first forest industry person to speak, I was the first industrialist to ever speak to that group.

We're going to disagree—there are parts of the Green Party policy that we think are excellent and parts that we think are wrong. But once you establish that you live in the same universe, that you all more or less get the same set of facts, you're really trying to figure out, "Okay, how do we solve the problems between us, and what's the practical course?" The conversations get easier. They get hard sometimes; various social values have to be reconciled. But once you can define the common value base and the common concern base that you're coming from, it gets a lot easier.

What are the most important things you think Canada needs to do for the future of the forest industry?

We have to continue to accelerate our transformation into the greenest forest industry in the world. What does the world care about with forest products? They care that the forests are regenerated properly and that there's no deforestation and that's not a difficulty for us because the forests are publicly owned. They care that the forests are harvested with care and that there's outside scrutiny with forest practices, which is why we do certification. They care about biodiversity and endangered species and there we're doing well, certainly much better than the illegal loggers and the clear-cutters. But we have challenges around endangered species, around the caribou and some of the songbirds. They care about waste, and we've committed to accepting every piece of scrap paper people want to give us for recycling. We've reduced what we send to the landfill by more than 40 percent. We're pretty well up to using 98 percent of the tree. But our recycling industry isn't near where we want it, mostly because we just can't get the post-consumer fibre.

People care about air pollution. They want a product that wasn't produced with air pollution. We've improved 50 percent in that area, but we're still not at the acceptable level. They care about water pollution, and we're close to where we think we should be; we're not finished but most of the toxins that we used to emit are now at non-detectable levels. They care about green energy, and we're about 60 percent green energy in our mills, but we can and will get to 100 percent. So it's that transformation. It's not the only thing. And we have to become more efficient and productive. Some of these things will help that but some won't, they won't detract from it. So there are other things to be done. But in the end, in a world of fierce competition, we export to the global marketplace. You have to be able to differentiate your product or you're just going to be selling at the lowest possible price. Environmental differentiation is what we're betting on.

There's a phrase in one of your articles that really intrigued me: "Building environmental progress into the economy." The old model of thinking of considered environment and economy is that they're conflicting values. And sometimes there are conflicts. You've got a town in which the mill is the source of employment and their wood supply comes from critical caribou habitat. You can say that environment and economy and social concerns don't conflict, but sometimes they do. People who lose their mill basically lose their life savings and social disruption goes up, and yet no one wants to lose critical habitat for such an important species.

So sometimes there's conflict, but much more often there's a convergence of economic and environmental interests. There's efficiency convergence—when we get off fossil fuels, we are not held hostage to what the oil cartels want to do to the western economy, we're not held hostage to even Canadian energy supplies that can get pricey. If we can burn waste, and most of our green fuel used to be garbage, we can be waste-based and more economically independent of that cost shock. If we can gain market advantage by giving people what they want, which is products that don't make consumers feel guilty, we can have market advantage. If we can run low-waste facilities and use every part of the tree—take out the logs for lumber, use the chips for paper, use the sawdust for plywood, use the bark for fuel—it's very economic and at the same time environmental. So there's an economic aspect of environmentalism. But I don't want to say it's only that, even though my job is as an industrial leader, because there's an underlay of values, of things that humans care about. So even though there can be economic advantage and we're glad to have economic advantage from environmentalism, there's also a moral and human imperative of living in this world with others in a way that's respectful of others and respectful of future generations.

When we have these conversations with our employees, when we have them around the boardroom table with CEOs, that sense of caring is also in the room, and it means something even to crusty old CEOs to go home and look their son or their daughter in the eye and be able to tell them what they're doing. People care about their integrity as well as their pocketbook.

▲ **You've worked in public policy. If somebody put you in change of Canada's forest policies, what would you want to see happen?**

I'd like to see the tax structure changed to make it more attractive to invest in Canada's forest industry. I'd like to see the monopoly powers that railways have to overcharge our industry—because we mostly have single lines that charge us too much and give us bad service—so I'd like to see the railway's monopoly powers bust up. I'd like to see greater support to accelerate our transformation to greenness. I'd like to see the government better recognize the carbon-sequestering properties of wood in their policies for procurement and building. And I'd like to see both national and provincial governments work more with unions and First Nations and local communities and shareholders of the companies to better shape the vision for where we're going and then create more circumstances for getting there. And, finally, as a guy who has to compete for a living, I'd like to see the government shine the spotlight on the bad actors. I'd like to see illegal fibre made the pariah of the world global trading system. I'd like to see logs coming from places that have deforestation have a taint of shame in them, that they won't be bought. I'd like to see governments insisting that all procurement be only for product that comes from certified forest. It's good to encourage the good, but I think there should also be a strong discouragement of the bad practices.

How do you feel about trees?

I certainly have a deep and wondrous affection for trees and forests at all stages of their life. I hang out in some very mature forests that have that special quality of a climax forest. Because I'm in southern Ontario for some of my canoe tripping and hiking in relatively young forests, there's a peace and gorgeousness to them too. And I actually feel quite at ease that I work in an industry that both harvests and replants forests.

Watching humanity live in cities of asphalt and tar and concrete and steel and metal and plastic, I love the idea that people can be surrounded with natural products, with wood, with its inner beauty and its gorgeous feel.

I heard you're an "aggressive vegetarian." Can you explain that?

I've been a vegetarian since I was 16. I became a vegetarian as part of a teenage spiritual quest, which was stimulated by Henry David Thoreau and one of his books, I don't remember which one, it was a long time ago. He said if a person wanted to preserve the delicacy of their soul they shouldn't partake of meat, and so I decided as a 16-year-old that that's what I was going to do for myself.

But my fierceness has only come out as I understood the contribution of animal husbandry to deforestation and climate change—because the single biggest cause of deforestation is clearing the lands for either soy beans to feed to the livestock or the livestock itself. But that's my hobby, not my job. I occasionally use the platform that my job gives me to take a shot at the beef industry.

Forest Products Association of Canada at www.fpac.ca

22
CHARLOTTE GILL
TREEPLANTER

There was only one character I wanted to add to the movie *The Green Chain* that I just couldn't squeeze into the script. I wanted to end the movie with a treeplanter. I'm friends with several former treeplanters and they all have amazing stories. They all describe treeplanting as something between a tribe and a religion. But that tribe always felt separate from the world I was creating.

But I knew when I started the podcast series—and this book—that I had to interview a treeplanter. The problem was there are so many people who have planted so many trees that I had no clue how to find the perfect planter—until I opened the literary journal *Vancouver Review* to discover a story that began, "We tumble out from pickup trucks like clothes from a dryer. Earth-stained on the thighs, the shoulders, around the waists with muddy bands, like grunge rings on the sides of a bathtub. Permadirt, we call it."

The words belonged to Charlotte Gill, a poet, author and proud treeplanter who went on to define the job perfectly. "Treeplanters— one word instead of two. Little trees plus human beings, two nouns that don't seem to want to come apart."

When I contacted Gill, I found out she was working on expanding her short essay "Eating Dirt" into a book, a memoir entitled *Spade Life*. I would have been disappointed if she wasn't.

I met Gill in April 2009 in Calgary, where she was writer-in-residence at the University of Calgary and sounding ever so slightly wistful about the fact that the job meant that this year, for the first time in a decade and a half, she wouldn't be heading back into the woods to plant trees.

When and how did you start treeplanting?

Many, many years ago actually. I've been planting trees for, I think it's 18 years now. It's kind of hard to tell sometimes because the years sort of bond together. But I started planting trees in Ontario because I lived in a house where other treeplanters lived and I had just moved to Canada from northern New York State, which is where I grew up and I really had no sense of how large Canada was at all. I knew it had 10 provinces and some territories, but I had no conception of how gigantic this country is, and how empty it is in a way. So they started talking about treeplanting and I imagined, I think, something really pastoral and Johnny Appleseed out of New England. And they said it's almost military with how tough it is and how austere it is.

And I think there was something about being outdoors and having a summer student job that allowed you to have so much freedom and not very much supervision that really appealed. So I went treeplanting that first year—I think it was 1989 or 1990—and I was hooked. It was like some kind of addictive thing in the air. I've heard many other treeplanters say similar things about going treeplanting for the first time. There was really no looking back. It's like a fold in their lives.

What hooked you on it?

I think it was twofold. A lot of it had to do with the geography of Canada. It really just blew my hair back. I went out to my first treeplanting contract north of Thunder Bay, and I got on a bus and I rode for 24 hours, something like that—it was a day and a night—and we were still in the same province. I got a glimpse of this beautiful boreal forest with these trees that were just a green curtain along the highway. And I thought, "Oh my god, this goes north all the way to the tundra." So that really appealed to me on a deep level, coming from the States where every 100 miles you run into a new town or some kind of hamlet anyway. Also I think there was something about working outdoors and having my hands in the dirt and doing something really tangible. At the end of the day I could look back and see what I'd done and say, "There are my 1,000 or 2,000 trees and I did that with my hands."

Apart from that, the people. The people were so wonderful. They came from everywhere. Some of them had already been planting for 10 or 20 years. They came from all over Ontario and east, and they all had really fascinating stories and reasons why they were also out planting trees. Many of them were artists and musicians and I was really attracted to that.

Were you a poet before you were a treeplanter?

Those two things kind of went hand in hand. I was attracted to the idea of treeplanting because it's very seasonal work. You can work really hard in four to six months and then have a whole winter off to do other pursuits.

When did you start with the poetry?

Once I'd finished university and moved out to British Columbia where I was working, say, from the end of January sometimes until October, which left me the whole winter to kind of do what I wanted—travel and write. Writing, like treeplanting, is a really lonely discipline. You have to spend a lot of time by yourself and believe that in piling up really small gestures, you'll actually achieve something much bigger than what it feels like in the moment. So I think just doing it, planting trees every day, one after the other after the other, it kind of gave me the sort of visceral belief that I could do it with sentences.

Have you gone back to the first place you planted to see what it looks like?

No, I don't even know where I was. The first time I was 19 years old. I couldn't even point to a map now and say where it was. I think that's a really common experience with treeplanters. A lot of the places we go, they don't even exist on a map. You could probably roughly ballpark it from looking at satellite photos. But those trees now would be 20 years old. They'd probably be quite impressive, to me anyway.

▲ **That's why I was wondering if you'd seen any.**
I've seen some clear-cuts that I've planted just on the southern end of the Great Bear Rainforest, Seymour Inlet. As a treeplanter you don't get to see it very often because you come, you plant the trees and then you leave.

I was really amazed to see how a lot of them were growing and to have such a physical memory of actually walking every step of that land, foot by foot. They were all growing pretty vigorously. I think now everyone in the business has got a lot better at planting trees well. In the beginning a lot of them died. Certainly being a treeplanter you know a lot of what you do won't end up surviving for whatever reasons. But trees are really good at growing. In British Columbia, on the coast, you can almost just throw one in the ground and it will grow by itself.

▲ **Can you talk a bit about the culture of treeplanting and what kind of people it attracts?**
In the course of writing this book, I really set out with this intention. I wanted to find out what that thing is, what that essence is that makes us all sort of similar. On the one hand, I found it really easy to do. I think that treeplanters all have this very furious way of being in the world. We are all very in-the-moment people in some way, each in our different way, perhaps. And treeplanters have this phenomenal athleticism. They're just incredibly fit people. You would never guess from watching somebody walk down the sidewalk, you would never guess what they can do on the side of a mountain. There are people who can climb like spiders and

carry enormous weights and get over gigantic obstacles that I think a lot of people would feel intimidated to even walk in their sneakers, carrying nothing—just incredible agility.

On the other hand, in the course of writing, I figured that there weren't a ton of things that really connected us culturally other than the fact of the job itself, which is so intense and so sweaty. And that maybe it was the job itself that made us so close in a way, especially people who have done it a long time. Really, it's changed us over the years.

▲ **The friends I've had who've done it always describe it as really tribal.**

It is really tribal. I think on some level there must be something genetic in that. I think there is something that happens to human beings when they're in a position of physical adversity—the natural tendency is to bond. And I think there's also something about performing that repetitive motion, everyone doing the same motion in a way that is very equalizing. There's some hierarchy in the treeplanting business, but not very much. There's a tremendous sense of equality. No matter where you've come from, no matter if you're a man or a woman, we're all doing the same work. And so that is liberating in a way.

▲ **When's the first time you saw a clear-cut?**

I saw clear-cuts as often as anybody else would—on television, on posters. I think they really disgusted me in the beginning. I had never seen anything like it. I thought it was carnage.

I had a sense of what a clear-cut was, but I didn't experience the scale of it until I actually stepped inside of one. At the same time I was blown away by how vast it was. In the same way that somebody might feel completely staggered by walking around on the sidewalks of Mumbai and experiencing the crush of all these people. Or walking on a glacier in Antarctica with it being just completely plain and white. You can see it in photos, but it's different to be there.

I think a clear-cut is just grey and spiky and it has a ton of texture, but at the same time is completely monotonous, like a parking lot or—a parking lot is probably a good analogy. It feels like that. It has that kind of loneliness to it.

▲ **You said it looked like carnage—has that changed for you now?**
Yes and no. I think that in some way treeplanters maybe have not been favourites of environmentalists in the same way that loggers aren't. We are the thin end of the whip in a way. We're the tail end of the business. We work in a clear-cut, we do it every day, we get inured to it—looking at stumps and slash and all that. But I don't think that can be helped. I think that's just part of adaptation; we all get used to what we're looking at. People in offices get used to working under fluorescent lights. And that's our workplace.

But the more I've worked in clear-cuts, the more I've thought about what they mean to us living on the planet. Clear-cuts are a collision site between two creatures. On the one hand, you have trees, which are ancient sometimes, 1,000 years old. But the forest they live in has been potentially building for 10,000 to 15,000 years. These creatures have a scale in time that is so completely beyond what we can fathom because we have these brief, barely 100-year lives.

On the other hand, you have humans and our appetites. That's probably why we're still around, because we have appetites. So you have something very ancient on one side and this very speedy reaping of resources on the other side.

A forest just wants to build, it wants to make more of itself. It can't move, which is probably its downfall.

▲ **What kind of trees have you planted?**
It depends on the region—I mean, dozens of different species of

tree. People often say to me, "What's wrong with a clear-cut?" What's wrong with replanting in a clear-cut is that you are planting a monoculture. And a lot of the time that's been the case, especially in the more interior parts of Canada.

But on the coast, I would say that sometimes I'm planting as many as six to seven to eight kinds of tree at one time. The idea is to try to replicate what was there before. Certainly in a climax forest that's an incredibly diverse amount of species.

I'm assuming that's a newer concept.
Not so much, actually. When I started planting trees in British Columbia, that was about the mid-1990s, people were doing it then. That was in first-cut ancient forests. Now we're seeing something different, I think, which is second-growth logging. There are mostly Douglas fir forests, which grow in monoculture. So we're planting mostly Douglas fir, if not 100 percent Douglas fir.

When is the last time you were out treeplanting?
Last year. Probably around this time.

And you're not going back this year?
That's right. I'm not going back this year. But I miss it.

Planning to go out next year?
I would be shocked if I didn't go out next year.

How come?
I love it. I love the outdoors. I love the landscape of British Columbia. I work almost exclusively on the coast now, on Vancouver Island and what was formally called the Midcoast Timber Supply Area, now known as the Great Bear Rainforest.

There is something about those old forests that is so awe-inspiring, and although I'm in the remains of an old-growth forest, I'm putting my hands into soil that is potentially thousands of years old and it's a foot to two feet deep [0.5 metres] and it's so dense with life. I think scientists are only fully coming to understand what kind of communities grow in the soil. There are crustaceans in there. There are shrimp in the soil!

On some level, we as treeplanters sort of feel that in our fingers. It's much different than planting in something that's been logged for the second time. The quality of it is completely different. I don't think I would ever be able to stop going there.

You were saying your guy's a treeplanter.

My husband is a treeplanter. And how well we understand each other has to do with the fact that we've seen each other in that environment. There is something about the job that really renders people quite naked. There's no social constructions. The mask only lasts for a couple of days and then there's so much fatigue that it comes off. There is something really attractive about interacting with people on that level where they just are what they are and that's okay.

Also there is the aspect of living a fairly itinerant existence for so long. Obviously if you're living with someone who's doing the same kind of job, then you don't have to leave them behind for half the year, which many treeplanters do.

What does it take to be a treeplanter?

The first thing people say all the time is that it's really hard work. I've seen men treeplanters who are 6 feet 5 inches and just rippling with muscle, and I've seen women treeplanters who are four feet tall and 100 pounds and there is something in their minds that just makes them want to work hard. So I would say that's the common thing. It's more a mental game than a physical one. And probably athletes would say something similar.

▲ **But there's got to be a mindset as well that goes beyond just your physical shape.**

There are things that happen treeplanting that would just never happen anywhere else. There are obviously all kinds of the stereotypically wild behaviours that treeplanters are famous for—the crazy woodland parties and huge bonfires. That's all true. But there's this sense that we are out in the woods and all we have is each other. And I like that feeling. It's this sense of that other person over there. Although they're 100 metres away, they really matter in this way that maybe we take for granted when we're walking around on a city sidewalk.

▲ **Is there anything you're exploring in writing this book that I should be asking about?**

When I got about, say, a million trees into my treeplanting career, it hit me that we spend a lot of time in the woods, but we don't really know it that well as treeplanters sometimes. We don't know the names for things because we're so focused, we have our heads down all the time. And sometimes we pick our heads up and say, "Wow, look what we're standing in!"

So I started to do some investigating into what it means, and I think treeplanters have really great stories to tell, mostly because the places where we work, we're walking in the loggers' footsteps. We see everything that they do. We go everywhere that they go. And of course this country is huge and what gets logged is kind of invisible to us living at home because it's hidden behind crazy mazes of logging roads and it's locked behind gates. And so much of it is hidden, a lot of it you just can't get to it. There are forests in British Columbia that would probably take a week of travel to get to. So there was something about revealing that hidden side of what it's like to work in the woods that really appealed to me. It changed my mind in a lot of ways.

▲ About?

I think all treeplanters know on some level that what they're doing is beneficial in some ways and it's also very complicated in others. The logging business is so arcane. You could probably get a Ph.D. just studying how it works and all the various formulae that are used to calculate cuts and stumpage fees and stuff like that, the business as well as the forestry science. That's hard to live with sometimes because everyone in the world somehow thinks of treeplanting as a very benign human activity. It's complicated for us in the sense that we sometimes feel like PR instruments. And that can be difficult because we know that sometimes treeplanting doesn't exactly do what the companies that employ us say it does, what they tell the world that it does.

In the end, I think it would be fair to ask, "What would the world be like, what would it look like without all the trees that have been planted in the past 30 or 40 years?" Probably much emptier. It had to be done by somebody.

▲ You've said you've planted more than a million trees. Have you calculated?

I lost track about eight years ago. I think it's probably around 1.5 million. But this is not really that much. I know people who planted around 4 million. They started planting trees back in the 1970s when it had just been invented by Dutch hippies.

▲ How many trees are there in a forest? To me a million trees sounds like several forests.

Well, it's a forest for sure. But these are little trees, and if you think about how long it will take them to grow to replace what was cut down—certainly a tree that's been growing for 500 years, you might have to leave those little seedlings for another 500 years to get them that big. Maybe it will never happen. I calculated what a million trees is. It's around 600 Manhattan city blocks.

▲ If somebody put you in charge of forest policy for BC, what would you want to do?

That's a big, big job. I don't envy that person. I have a feeling that environmental groups are pretty clear on what they want, what they think are good ideas. And I would say their ideas are probably the best. They spend a lot of time thinking about it and acting on it. I'm not an activist, and I'm not a journalist, but I think the idea of stopping old-growth logging is probably a really good idea, as well as the banning of raw log exports. And probably from my own perspective, and this is incredibly scientific, there is something vastly different about a second-growth plantation compared to an intact old-growth forest.

But I wonder what will happen once we start really cutting those second-growth trees down. I think on some level we can have sustainable practices, but in a way we need to match those practices with sustainable appetites. Where is this wood going?

I realize we need wood, and clear-cuts will probably never go away. But what are we using it for? If someone sneezes on the other side of the world and uses a Kleenex made out of ancient timber and throws it away, is that an appropriate use of something that took so long to make?

▲ How do you feel about trees?

They inspire me. They're wonderful creatures. We as human beings, we walk around on a really horizontal plane and trees occupy three dimensions on a scale that I think is lost to us.

We're only beginning to understand. They have these entire canopy ecosystems living in them. And they have whole worlds under the soil. They eat light. All they want to do is build. I think they're amazing creatures.

Charlotte Gill at www.charlottegill.com

THE FIRST LINK IN 1
THE SCREENPLAY

The Green Chain film received its world premi
Film Festival in Montreal. The film won the "   ⌐gat
Award" (the Award of the City El Prat de Llopregat) at the 15th
Annual Festival Internacional de Cinema del Medi Ambient (FICMA
2008) in Barcelona. The script was a finalist for a 2008 Writer's
Guild of Canada Canadian Screenwriting Award for "Best Feature
Screenplay." *The Green Chain* has been released in theatres across
Canada and broadcast on TMN and Movie Central and is available
on DVD and legal download sites around the world. For more on the
film, please visit www.thegreenchain.com.

The original screenplay was written with specific line breaks to
indicate rhythm, pace and style. In order to be able to include the full
script here—and to create a cleaner read—I've chosen to go with
more traditional line breaks.

FADE IN:

EXT. FOREST—DAY

An endless sea of ancient trees as THE GREEN CHAIN SONG plays and we PULL BACK to reveal
The hydro wires in front of the forest and PULL BACK to reveal—
Pristine forest and PULL FARTHER BACK to reveal—
SLASH and the music begins to be replaced by the music of loggers at work . . . Saws chewing through trees, branches flying, a danglehead processor eating through a forest.

EXT. PROCESSOR—DAY

We find the machine that's turning trees into lumber and, inside it, a rugged, likeable man, **BEN HOLM.**

EXT. PROCESSOR—DAY
Gearing down.

INT. PROCESSOR—DAY

                      BEN

  I love trees.

Ben opens door, steps out of machine.

EXT. PROCESSOR—MOMENTS LATER

Ben indicates the machine.

BEN (CONT'D)

Beautiful machine, eh?

Call this a danglehead processor.

Once the tree's cut, this baby will limb it and buck the logs at 16 feet a second. Then we just have to stack it and haul it out to the landing.

You have any idea how long it used to take to do all that? How many men it used to take?

The falling—that's always been fast.

Not all that safe mind you—

Don't know too many old timers still have all ten fingers.

My old man used to say if you had all your fingers you probably weren't very good. Least not very fast.

My old man had seven.

Now he was fast.

Could take that tree down as fast as any machine.

Loved to watch him work.

He took a few competitions too.

Even went down to the city to compete a few times.

Won a Championship. Back when I was kid . . .

And people cheered for him.

People cheering for loggers.

City people.

Hard to believe now . . .

Felling, logrolling, climbing.

The old man couldn't log roll for shit—

Not much chance to practise up here.

But at the big competitions he'd do that too, 'cause if he was gonna enter that competition he was gonna make sure he was in every event.

So he tried logrolling. One, two, splash.

Told me years later he was sure he could have made it a few more seconds if he hadn't been sober.

In front of trees (and machine).

BEN (CONT'D)

City people.

Yeah, when I was a kid they actually knew where wood came from.

You know the saying, money doesn't grow on trees? Old man used to say . . . Money is trees.

That cedar there—

The one we just took down.

Two hundred and fifty bucks. Easy.

Not bad, eh? That'll help pay for my kid's books.

Jason's at university. Studying silviculture. Silviculture.

I can't believe I'm paying for—

You should talk to him.

He'll tell you all about trees.

When he was younger, I thought maybe—

But it's not like there's any work here.

I been on part time six months. And I'm lucky.

Yeah. Lucky.

Wife had to go back to work so we could keep the house—

But at least she found something. Even if it's serving burgers.

And now our kid's studying silviculture.

So he can come back for Christmas.

Tell me what we're doing wrong.

Still like to take some of the trees down myself.

Sometimes I'll even go out and practise with the axe.

Couldn't do competition though—

Don't have the swing for it.

Had an accident when I was a kid.

Working on the green chain.

Wasn't looking. Or somebody wasn't looking.

Arm got caught.

He displays his arm for the camera.

BEN (CONT'D)

Doctor said I was lucky it didn't shatter. Just broke it . . . Clean. So I never had a good swing. But now . . . Now I don't need a good swing.

Mostly use a chainsaw now.

Imagine having one of these machines in the old days—

Before the fucking environme'talists—

Sorry, don't quote me on that, okay?

Before the environme'talists started in—

I can handle vegetarians. My son's a vegetarian. Lentil this and tofu that.

You don't wanna eat meat that's your choice. More steak for me.

But trees?

What do they think their protest signs are printed on?

. . . Few weeks ago ran into a protester. He was lost. Typical.

Looking for the forest he wants to save and can't find it. Sees my chainsaw and says I'm a murderer . . .

Is that a smart thing to say to a man with a chainsaw?

Then he asks me where the protest is.

Went to take a leak and got lost.

Went to take a toke is more like it.

So I ask him what he lives in? What his house is made of. Then he hands me a pamphlet. "What's this?" I say. "Plastic?" He doesn't laugh.

Then I tell him I got a great idea so we don't have to cut down any more trees.

He says, "Yeah."

Musta been stoned.

So I say, "Yeah. Here's how we do it. Here's how we stop
cutting trees. From now on we make all our houses out
of baby seals and whale bones."

He does not laugh. Not even a grin.

Not sure he knew I was kidding.

So I tell him the protesters are that way.

Ben points to the stand of trees.

BEN (CONT'D)

And he starts heading that way.

Protesters are that way.

Ben points in the other direction.

BEN (CONT'D)

They'll prob'ly find his body in the winter.

City people. Think trees are house pets.

They're trees. Lumber. Pulp. Sawdust. Money.

Stop looking at me like that. I'm just kidding.

I pointed him the right way.

Though God knows if that helped.

Yeah, sure, no problem, bud—

Feel free to climb up there, so I can't pay the mortgage.

Yeah, my pleasure. But call the bank for me, would'ja?

And could you feed my kid while you're at it? And pay
for his university? You'll like him—he's a vegetarian. He
hates me too.

Shoulda pointed him the wrong way.

See if the trees talk to him—

Tell him how to get home.

I heard of some kids—

Spike trees so they'll break your blade. Maybe kill you when it does. Toss sugar in your gas tank. Not here—

But it happened up north.

Anyone does that to my baby, then I'd get out the axe . . .

See how good my swing is . . .

How can anyone do that?

You wanna protest? Fine. You got your right to be an asshole. I got mine. That's what we call democracy. But how can anyone—

That machine's my livelihood. It's my life.

Bad enough they keep coming up with new places we can't cut and new reasons we can't cut there . . .

Old-growth forest. Know what that means? Big trees. Know how they got so big? Fire hasn't got 'em yet. Or beetles. It's a miracle.

Know how many trees we lose to fire every year?

Know how many we lost this summer?

Don't see Greenpeace protesting that.

Wonder how many of these asshole protesters forget to put out their campfire—

Or toss a joint—

And take out more trees with a spark than I could in a lifetime—

I'm not the enemy here.

I'm a logger.

My old man was a logger, his old man was a logger—

His old man fished.

But he lived in Norway so he had an excuse.

Used to be you could cut pretty much anywhere—

Now you need a special chainsaw just to make it through the red tape . . .

I got this beautiful machine.

And they want to tell me I can't go out there—

Can't cut anything.

Damn environme'talists.

Buncha asshole university kids never had to make a cent in their lives come up from the city in the SUVs their parents bought 'em and tell us how we're supposed to live.

And the government buys it, 'cause it's the city that calls the shots. So every day there's less trees to cut.

Maybe I shoulda pointed him the wrong way. Woulda served him right.

If he doesn't know his way around the forest, where does he get off saying he's the one who's talking for it?

I grew up here.

And these shits should be cheering for me—

Just like they cheered for my old man back in those competitions—

Because this forest is paying their bills too. Paying their taxes. But they don't see that anymore.

They don't wanna see that.

Like they don't wanna see hamburgers come from cows.

They just wanna hand out their pamphlets—

And read their newspapers—

And forget where all that paper comes from.

This is my home.

This forest . . . it's my life.

These are my life . . .

I love trees.

(beat)

You good? I gotta get back to work.

Ben returns to his machine. As we leave his forest, we hear the processor rev up again. The **sound** *of wood being SLICED, chips flying—*

INT. A GREY ROOM—DAY

The open, friendly face of a vibrant grey-haired granny, **ABIGAIL EDWARDS.**

> INTERVIEWER
> How do you feel about trees?

> ABIGAIL
> I love trees.
> I always loved trees.
> And this—
> This is worth it.
> They asked if I was willing to go to jail.
> I said, that's what civil disobedience means.
> People forget that.
> I saw an abortion protester on TV.
> They sent him to jail.
> He said, "You can't send me to jail. It's civil disobedience."
> I've heard some of the kids I'm out there with say it too.
> "They can't send me to jail. This is civil disobedience."
> And I tell them—
> Civil disobedience means breaking the law because you don't believe in that law. And it means you believe in something so strongly, you're willing to go to jail for it. You're willing to go to jail to change that law.
> So that's what I said to them—
> You want to send me to jail—

Send me to jail.

Because if you don't, I'm going back out there.

And you can arrest me again tomorrow.

INTERVIEWER

How did you get involved?

ABIGAIL

I used to teach science. Before I retired. Photosynthesis.
Ecosystems. How we need the trees and the plants—
So we can breathe.
When the city tried to clear-cut our park for a
highway—
Back in the sixties—
I was there. Protesting. Organizing. Raising hell.
I was an environmentalist before it was a word. Back
then they said I was too young. Didn't understand the
way things work. Now they're saying I'm too old. That
I should stay home.
Mind my own business. I'm not supposed to care. I hear
it every day. "Go home grandma."
Not from my grandchildren.
From other people who assume I'm a grandmother just
by looking at me.
The kids are mostly polite about it. And they always
share their coffee. Or food. Or cigarettes. One boy even
offered to share his mushrooms. I passed. I appreciated
the gesture though. All these pierced young things—
Really quite polite when you get to know them. If you
talk to them. Not at them.
There are bulldozers out there building a logging road.
And these kids are out there trying to stop them.
They're laying in front of bulldozers.

And when I showed up one of the girls with spiky purple hair and a ring through her nose said to me—
"You don't want to be here, ma'am."
She called me ma'am.
Not grandma. "Ma'am."
And she said—
"You don't want to be here.
We're lying down in front of the bulldozers."
And I said—
"You think I'm too old to lie down?
I can lie down all day."
When are you supposed to stop caring?
Sixty? Sixty-five? Seventy?
Is there a mandatory age when you're supposed to say to hell with it all . . .
After you retire? If you can afford to retire.
I own my house. Ever since Martin died. It's mine. Insurance cleared the last of our mortgage. And my pension's enough for me. So what does that mean? I'm supposed to sit in a rocking chair and knit and watch bad TV?

INTERVIEWER

How far will you go?

ABIGAIL

There are other women out there . . .
The women on Salt Spring—
They wanted to save the forest. So they took their clothes off. Posed for a calendar. Naked.
That got people's attention.
Nothing makes people take your cause seriously like taking your clothes off. Nothing like a woman's breasts

to get the media interested in "the issues."

Do you know how they stopped the seal hunt?

Brigitte Bardot. Greenpeace flew out Brigitte Bardot.

And every TV station and radio station and newspaper on the planet sent someone out to interview her. And they asked the spokesman from Greenpeace, "What's she doing here? What does she know about baby seals? She's not a scientific expert."

And the man from Greenpeace smiled and said, "Would all of you be here to interview our scientific expert?"

That's what you call a rhetorical question . . .

And suddenly the whole world knew about the seal hunt.

I'm a grandmother. I'm not supposed to be a shit disturber. At my age I'm not even supposed to say "shit."

But that's the point. That's why I'm here. Because I'm a grandmother. I have grandchildren. They have to live on this planet. And—God willing—so will their grandchildren. And I'd like them to live on a planet with thousand-year-old trees. Or two-thousand-year-old trees. And air. Remember photosynthesis? And forests. Not tree farms. Forests.

I'm an environmentalist.

And that makes me a criminal.

You know what criminal is? Criminal is clear-cutting so many trees that the scar on the earth is visible from the Space Shuttle.

Criminal is attacking a forest filled with thousand-year-old Douglas firs.

Criminal is letting corporations decide what's good for us.

Criminal is arresting me and my friends—

And letting the loggers and the corporations go free.

And if I go to jail—
Maybe my granddaughter won't have to lay down in front of a bulldozer one day.
So no—
I'm not going to stop. And maybe—
just maybe—
some other grandmothers out there will start to think—
Saving the forests—
Fighting for the future—
This is better than knitting—
Or watching reality television.
I protested. I broke the law.
I'm guilty. I plead guilty.
My crime?
. . . I love trees.

A brutally clear-cut forest morphs into the charred remains of a burnt-out forest and—

EXT. FOREST—DAY

A stand of blackened trees and stumps. As we close in on the burnt-out husks we see **BRETT HALL.**

                    INTERVIEWER
How do you feel about trees?

                        BRETT
I love trees.
I always loved trees.
I been fighting fires since I was 16.

Was my first summer job.

Then I joined the company.

Got a job on the green chain.

Was a grader—until a few years ago.

They didn't call it firing. "Rationalization."

Didn't matter what they called it.

No work. No money—two kids.

So I signed up to fight fires. Not like you forget how.

'Licia says it's the craziest thing anybody could do.

But it's gotta be done.

Every summer some camper drives up from the city, toasts up some marshmallows and after a few dozen beers figures the fire'll just put itself out.

Or they toss a few rocks on it.

And it looks like it's gone.

Fire's a tricky thing. It can hide. It does hide. You can hit it with all the foam—or all the water—in the world. Looks like all the oxygen's gone. Nothing left to burn. But if it's not ready to die some spark'll find a way through, find some more tinder and—

(snaps fingers)

BRETT (CONT'D)

I'm back in business.

When I was a kid I wanted to be a jumper.

I remember I thought—

They look like superheroes in those outfits.

Look like they're flying. But they know it so . . .

If I was single maybe—

But I never was single for long.

And the mill was steady.

I'd always have a paycheque . . . Yeah.

Sometimes it's lightning starts it.

Don't get too much lightning up here but when we do—

See, with your man-made fire it's all spreading out from one place. Think about it, eh. Campfire gets out of control and yeah, it can spread, it can jump, it can fly—you ever seen fire fly from treetop to treetop?

You see that—you really have to remind yourself it's not alive. It sure looks alive. Sure seems alive. It eats, it grows. Just never sleeps.

Cavemen thought it was sacred.

Scared the hell out of 'em . . .

Scares the hell out of us too.

Still seen people, people like Dale—friend from school . . .

Fire's a tricky thing.

Lightning. Now that's nasty.

See the thing with lightning is maybe you get a dozen strikes. Maybe you get a hundred. And then they have to call in everybody.

If there's a big lightning strike and you know how to go in you're in. That's how I got in when I was 16. Not that I knew how—but I was willing.

And I was big enough to carry the gear. That's all it took. See that's the thing. Now every job needs training . . .

The jumpers—most of them have university.

What the hell do you need university for jumping into a fire? All you need is balls.

But that's not how the government sees it.

Now you need training for everything.

You should see how the ladies look at the jumpers.

Like they're hockey stars. And the rest of us? Doesn't matter that we walked right into the fire. Nowhere near that kind of equipment, that kind of protection. We're not even there. And we're fighting for our forests—our home. Our jobs.

The jumpers, they just fly in—wherever—
and fly out . . .

I shoulda been a jumper.

Lotta kids, that's their first job in the forest—putting out a fire. In a bad summer you can work every day. Every day. For as long as you can keep standing.

Money's not bad either. Not mill money. But not many mill jobs around. And most of 'em you need university for that too. Or college or . . . To chop wood. Because of the equipment—the computers. What kind of world . . .

Sometimes the fires—sometimes some asshole comes up from the city, doesn't even think about it. Doesn't even try to put out their campfire.

Police catch 'em sometimes. Don't know how—but they catch 'em. I guess they leave their trash there. You know, an envelope with their name on it or something. Or they just figure out what campsite it all started at and who was there. Amazing, eh? That police could find that out. Look through a whole forest fire and find out exactly where it started. And when. And who started it. I can see how you'd need university for that.

Sometimes the loggers start the fires. Spark from a chainsaw or something.

A lot of fires start with a cigarette. Not from loggers. Loggers know better, eh?

Cigarette fires—people just aren't thinking.

When Dale died—when we were kids . . .

That was a cigarette fire. Who knows how they figured it out. But it was. They say it was.

At the funeral, Jenni—Dale's sister—
she and I were . . .
Well, that was a long time ago—
Jenni pulls out a pack of smokes, flashes it at me from
behind her Bible so no one else can see and says—

(whispers)

"Smoking can be dangerous to your health."
And I say, "Dale didn't even smoke."
And then we both start to lose it.
We're laughing so hard, but we're covering our faces—
So it looks like we're crying.
So we sneak out of church and everyone's looking at
us like . . .
They're all looking at us—
But they think we're crying and that makes it even
worse.
And when we're outside, in the lot, we finally pull it
together and Jenni looks at me.
I look at her. We stop laughing.
We're starting to get our funeral faces back on and then
she says . . . "Cigarette?"
And we both start laughing again.
And we both light up . . .
You mind?

Brett reaches for a smoke and proceeds to light up.

BRETT (CONT'D)
It wasn't the fire that got him exactly.
It was a tree.
Fire hit one of the branches up top and—

Summer before last there were no fires.

None. Lotta rain.

Good news for the mill.

Sure there was lightning, but whatever it hit didn't feel like sparking.

And there were campers. Maybe none of them felt like marshmallows. Too many carbs.

Or maybe just nobody smokes anymore.

I worked night shift at the Dairy Queen that summer.

Usually I'm in the forest—taking out a fire for twenty bucks an hour—but that summer . . .

I'm getting seven fifty, making sure all the kids and the tourists get their butterscotch sundaes no nuts and listening to some fourteen-year-old with a stick up his ass tell me I'm dripping too much chocolate on the counter when I dip the cones. Night shift.

So what am I supposed to do?

Shoulda told him where to stick his . . .

Shoulda—

I got two kids.

So I try not to drip.

Didn't even bother trying to switch to days.

Not like I had money to spend in the bar anyway.

I had friends couldn't find anything. If Dale was still around . . .

I guess maybe he'd be dipping cones with me.

Used to be a man could start work at 16 and know he had a job for as long as he wanted to keep working.

I got a wife—two kids—and I'm dipping cones. Nothing else to do. I could leave but—I don't know anybody anywhere else and I don't know what the hell I'd do when I got there. Met one jumper, Kennedy, lives in Vancouver—one of the only ones who ever talks to us "civilians." Said I could look him up if I was ever down there. But what do I do in Vancouver? Fight fires, grade lumber or dip cones?

Right.

So every day I'm putting on my little white ice cream suit, looking out at the forest wondering if this is gonna be the day.

We PULL BACK and pan the remains of the forest and—

INT. MOTEL ROOM—DAY

A megawatt smile belonging to a stunning female face that just has to be famous, **LEILA COLE**.

LEILA

"I love trees."

(beat, adjusts approach)

"I love trees."

(beat, another approach)

"I love trees."

(hesitant, questioning)

"I love trees."

Are you sure they want me to say this?

Isn't love—

You know—

I mean trees are great and everything—

And I want to help but—

Isn't this going to sound—

No, it's okay, I'll try it again.

"I love trees. And the reason I'm here today is because we have a unique opportunity to save this precious echo-system."

That doesn't sound right.

Eco-system? Make a note to check that, okay.

Someone's got to know. Echo. Eco. Just to be sure. I don't want to get it wrong. Echo-system? You're the environmental expert? Eco or echo?

INTERVIEWER (O.S.)

Eco.

LEILA

Thanks.

Maybe that's the Canadian pronunciation.

I know I've heard both. We'd better check.

"This isn't about me."

I'm not going to say that. It makes it sound like it's all about me. Who wrote that, my publicist? Did Joanne ask you to put that in? Just take it out.

Can't we just start with, "Thank you?" Thank them for coming? I mean, think about it. None of these reporters really wants to be here. No one wants to be here. They all had to fly out—or drive up—so maybe if I say that up front. Acknowledge it.

I mean, look at this place. Did you see where they put me? I've got the nicest room in town. The nicest room. And it's in a motel. The last time I was in a motel I was getting chased by a slasher.

And I got to go back to my trailer when they yelled "cut."

Do you think they should yell "cut" in a slasher film?
I think in a slasher film "cut" should mean roll.

Not stop.

The only time I've been anywhere this—

The whole place looks like the set for a slasher film.
Or a vampire movie. Or maybe one of those plague
things. The buildings are cheap. The cars are cheap. The
stores—have you looked at the stores? Half of them are
closed. Even the ones that are open . . . Everyone looks
depressed. Like the vampires are already picking them
off one by one.

Somebody should film a vampire movie here. It'd be
perfect. And here I am—on my one weekend off in a
month—in the only town left on the planet with no
Starbucks. Not even a McDonald's. A Dairy Queen.
That's it. A Dairy Queen. And we're still an hour from
the blockade. And now I'm supposed to learn this. And
I should probably memorize it too. So it doesn't sound
like—

I guess I don't have to memorize it. Nobody's going to
pay attention. Nobody wants to hear what I have to say.
The reporters don't. The kids on the blockade—they
don't. They're risking their lives—lying down in front
of bulldozers. Sitting in trees a thousand feet high. And
a hundred years old.

They know all about the trees.

I'm here because everybody knows the TV cameras will
follow me. Like you did. And they'll take my picture.
Hopefully in front of the blockade—standing with the
protesters—but probably in town—when I'm eating
lunch, or walking into my motel room.

And if we're lucky we'll end up on *Entertainment
Tonight*—or in *People* or *Entertainment Weekly*—and

they'll do a story about the forest—and the ancient trees—and the kids risking their lives to save the trees—and maybe that old lady. The one who got arrested. But probably they'll just show the picture—and say this is where I am this week. Maybe talk about what I'm wearing. Or the movie. Or Tommy and me. And how he's not here. Not even mention the trees. Make it sound like I lost it again.

And then my mom will call because one of her friends will show her a tabloid story.

Or she'll be watching the news.

And it'll talk about how I'm buying the town. Or runningoff with a mountain man . . . Or a cameraman. Or joining a cult. Or in rehab.

So what's the point of making the speech if no one's going to listen? Can't I just stand there? Thank everyone for coming. Offer my support. Ask people to send money? Write them a cheque—and get back on the helicopter? And go—somewhere?

I thought it would be pretty. Have you seen the pictures? It's beautiful in the pictures.

Why does anyone even live here? There's nothing here. Nothing.

Did you see those houses when we flew in? Did you see where the fire was? When we flew in they showed me where the fire was. Shit. An entire forest gone—because somebody forgot to put out their cigarette—a whole forest—it went on forever—the fire—and now they want to cut more trees?

Where are the animals supposed to live?

I thought it would be pretty. Fun.

Not—

It's like a ghost town.

Can you imagine if Tommy'd come?

Can you imagine if he saw the room?

Ate the food?

Did you see the look I got in that place when I asked them to forget the bun? And when I asked for salad instead of fries?

It was like she was mad at me for being there.

It's not like we didn't pay for it.

And all the people there—

The way they looked at me. It was like—

You don't think they're mad at me, do you? You don't think they're—shit, they're probably loggers. And they know why I'm here. You don't think she spit in my salad, do you? I gave her almost a hundred bucks for a tip. You don't think she spit in my salad? Shit. You could definitely do a vampire movie here. Or a slasher film. They already have the chainsaws . . .

Come on, that was funny.

Maybe I should say that?

Make a joke about a chainsaw massacre . . .

Just because you're not laughing . . .

Don't tell me it's any worse than "I love trees."

I know it's important. I do. That's why I'm here. But no matter what I say—if I say anything at all about trees—they'll just make me look—you know.

I want to help. I do. That's why I sent money. It wasn't just the tax thing. I could put my money anywhere. I could donate to AIDS. Or children. Or—children with AIDS. Anything.

But I kept thinking . . .

Without the forests—without nature—

Have you read what they're doing to the rainforest in Brazil? Did you read that article about Sting? And how he went to Brazil. And bought up all that land?

Maybe we could do that here. Maybe we should . . . Just buy up all the trees—and those guys could take their chainsaws somewhere else.

Cut down other trees somewhere.

Or maybe plant trees instead. Do you know I read on the Internet that some of these trees can sing?

It's true. Singing trees. Who knows what's out there? And we'll never know if we keep cutting it all down. Destroying the echo-system or the eco-system or the—

The planet. That's it. I'm just gonna say "planet."

They should just take the pictures of me—but talk to someone else. I know the protesters want me out here—they brought me out here—but now that I'm here—what am I supposed to say? I love trees?

They should talk to the kid living in the tree—

Or the old lady.

We're trying to option her story, you know—

About how she went to jail—

To save the trees.

I'd love to do that part. But we'd have to change the story—so she's not that old. Mid-twenties. And a single mom. So there's jeopardy there. Like *Erin Brockovich*. But with trees. It's a great story. Powerful. That's the kind of thing I want to do—

Tommy would hate this place.

I know. I should learn this thing. I should have read it on the plane. But I had scripts. I always read scripts on the plane. It's so relaxing. The only place the phone doesn't ring. I just feel like . . . There's no way to win. If I act like an expert everybody knows I'm just—acting. And if I admit I'm just here because I care, because I want to help, then it's, "What do you know about trees?"

What do I have to know?
It's simple, right.
Clear-cutting is wrong.
Cutting ancient forests is wrong.
Take your chainsaws somewhere else . . .
Can't I just say that?
Can't I just stand out there—
And do the photo ops—
And say that?
I'm not an expert.
I just care.
You know?
Isn't that enough?
Isn't it enough to just care?
Can't you just know something is wrong?
Even if you don't know how to fix it?
Can't you just want to help?
I'm sorry.
Let's try this again.
I'll make it work . . .
"I love trees."

EXT. FOREST—DAY

DYLAN HENDRIX—*a true believer in his early 20s—framed on a camcorder, one hundred feet up a cedar tree.*

                    DYLAN

I love trees.
I want to be like Butterfly.
You know who Butterfly is, right?
You've gotta know who Butterfly is.

(reverential)

Julia Butterfly Hill. She lived in a tree for over two years. Lived there. It was nearly two hundred feet up. Imagine. Two hundred feet up. The platform was six feet by eight feet. Imagine living in a space that's only six feet by eight feet. Anywhere. An apartment six feet by eight feet. Now imagine it's two hundred feet up in the air. I'm only a hundred feet up. And that's high enough. And I've only been here a week. And I'm not sure how long I can last—but it won't be two years. My platform's eight by eight.

He turns the camera to scan the platform, shooting everything he describes.

DYLAN (O.S.) (CONT'D)
And I have a cooler. And books. And warm clothes. And friends bringing me food. At night. When the loggers can't see them. Police can't catch them. And this digicam. Thanks for this. So I can talk to you—

He sets his camera into a tripod, comes into frame, then continues—

DYLAN (CONT'D)
Butterfly.
Her tree . . .
There were storms.
There was lightning.
She could have died.
All because she loved trees.

He lifts his copy of Julia Butterfly Hill's book.

DYLAN (CONT'D)

Some people said the loggers might try to kill her.
Chop down the tree.
Shoot her. Maybe it was just talk.
Maybe.

Dylan puts the book down, stares at the camera, considers.

DYLAN (CONT'D)

Maybe it wasn't.
Maybe they knew if she stayed up there long enough, people would start to listen.
She was talking for the trees.
And after she was up there awhile, she says she started to hear the tree talk to her.
I don't like Tolkien—all those Dwarfs and Elfs and Hoblits . . . Used to think they were called Hoblits before the movies came out.
Never got into that.
Didn't even like the movies. Fantasy never did it for me. Always liked science fiction better. Or comics. Stories about the future—because it's hopeful, you know—
The idea we have a future.
Even if it sucks, like the future in *The Matrix*.
Just the idea it's gonna last. It's hopeful, ya know.
But he had trees that talked. The Ents. I liked that he had trees that could talk. And walk. And fight back. I wish our trees could fight back. Some people say they can sing. Singing trees. I wish they could sing. I bet it would sound like whales. But sadder. Deeper.
Her tree was called Luna. Sounds like the God of the moon. Butterfly and Luna became like soulmates or something. At first people made fun of her. They did

stories in the papers and on TV and it was like she was
a big joke. Crazy hippie chick living in a tree. They figured she'd last until the first time it rained. If she lasted
that long. Maybe she'd fall out. Die.

It happens. A few tree sitters . . .

I mean if you fell from two hundred feet—

or even a hundred . . . But she stayed.

The first time I saw her was in a magazine. *The Utne
Reader*. Great magazine. You should buy it. It has stories
about people who are trying to make the world better,
you know.

People like Butterfly. People like the old lady who was
out here. The one who went to jail. People like me.

They had a picture of her—of Butterfly—smiling,
fearless, two hundred feet off the ground. She has this
beautiful long hair and in the picture it didn't look black
or brown it was more like . . . superhero blue. From
comics. Like Superman or Wonder Woman's hair. And
she was smiling. And the smile was kinda crazy. Like
someone who had a secret—like someone who'd seen
God. And knew God's best joke.

And I saw that picture.

And I read her name . . . Butterfly.

I saw her there. In the tree. And I thought . . .

It was like beyond thought. It was like all my breath
disappeared. I thought . . . She was as beautiful as—a
butterfly in a tree. Perfect.

And here she was saving trees.

I'd signed petitions. I marched. Wrote letters to
politicians. Form letters, but still, signed my name.
Gave money when I had some. Gave less when I didn't.
The idea that we still had ancient forests—think about
it . . . Ancient forests.

Not a hundred years old. Ancient. It's amazing. We've got forests that have been around—like the pyramids. Longer. They've got plants, bugs, maybe even animals we've never even seen.

I still think there's Sasquatch out there somewhere.

And we're cutting them down. For what? Toilet paper? *People* magazine?

The Utne Reader's on recycled paper. Just thought I should say.

And these forests—they're like the true wonders of the world. And some of them—they're only like a few hours away from my house. Not like Butterfly's forest.

And then I saw the old woman. Abigail. Saw them drag her away on TV. Saw that she went to jail. And then they had the concert—and you know who showed up to talk.

(amused and impressed)

Yeah. Can you believe it?

Did you see that dress she wore at the Grammies?

(still sounds impressed)

Can you believe she was here?

(now, not so impressed)

Like she knows anything about trees?

She couldn't even say "eco-system."

But it worked.

Suddenly this forest was famous too.

And the whole world knew we needed to save it—

The whole world was watching . . .

They figure Luna—Butterfly's tree—was a thousand years old. A thousand years old. And it was still alive—and they were going to kill it. To make toilet paper?

There are thousand-year-old trees here too. A thousand years. Ten centuries. Before TV. Before radio. Before Shakespeare. Before . . .

Think about it. What were people doing a thousand years ago? And it's still standing—a thousand years—it's part of the earth. We should respect it . . . We should worship it . . . Not . . .

And I thought—maybe writing letters—maybe that's not enough.

That old lady, Abigail, she didn't just write letters.

She put it on the line. And if she could do it . . . And if Butterfly could do it . . . So I came out here. And I volunteered. I thought . . . Maybe I could make a difference—remind people they're out here—these ancient trees. Remind the loggers what they're really cutting down. What they're killing. And it is killing.

They're alive.

Even if they don't talk to you. Or sing.

Butterfly saved a forest.

Maybe I can do that.

Be like her.

I don't want to be famous like her. I mean . . .

Not like that would suck or anything. Doing this is cool. Talking to you.

Is this really gonna be in a movie? I was just gonna put this online. But this. This thing you're doing. This is better.

EXT. TREE PLATFORM—NIGHT

Dylan—lit by stars and a lantern.

DYLAN

Getting arrested would be okay.
I don't want to get arrested—
But I could handle it.
I know people who got arrested on blockades.
They said it was no big.
So I could get arrested.
My parents would kill me.
But I could get arrested.
Maybe it'd help.
Maybe if more of us went to jail, more people would
pay attention—I've been here a week—no TV cameras.
No reporters. Just you.
It's not news anymore.
Its been done.
By Butterfly.
By other people.
And the *stars* have left. Now it's just us—
People you haven't heard of.
That's why these are great. So I can talk to you.
To the world.
Let people see what it's like.
If I last up here a year it'll be news.
If I get arrested, it'll be news.
If I die . . .
No reporters.
A few loggers—
Some called me names—
Swore at me—

Laughed—

One of them—this morning—he got out a chainsaw.
Revved it up—held it next to the tree—and for a minute
I thought—I'm a hundred feet up. I'm an hour from the
nearest real road. The nearest hospital . . .

You know the thing about the tree falling?

If my tree fell no one would know what happened.

Just the logger and me. And then I thought—

A hundred feet up. Just the logger will know.

And he held the chainsaw there for a minute—

Not even. Seemed like an hour.

Why would he do that?

I didn't do anything to him.

Then he stopped. Laughed.

"Just practising," he said.

"Just practising."

Killer, I thought. Killer! But I didn't say it.

A hundred feet up. And he had the saw.

Why can't he get another job, you know?

Planting trees or—

Another job.

Why does he have to kill them?

Why do we have to kill them?

Imagine living a thousand years and then—

A chainsaw.

And you're gone.

Like that.

And the logger—

He told me to look after myself. Said it was gonna
rain. Said it was gonna be an ugly night. And maybe, I
dunno—maybe it will be.

Radio said it could rain. I've got a radio too. Reception's
not bad at night.

Maybe he was trying to be nice. Or maybe—
Maybe he was trying to scare me.
Maybe if I'm up here long enough—
Maybe the tree will talk to me.
I hope so.
I had the camera.
I should have got it, pointed it at him—
But he had a chainsaw. I had a camera.
Ever play Rock, Paper, Scissors?

Dylan sticks out two fingers held together to indicate . . .

Saw.

He shows his thumb and forefingers in a circle to indicate . . .

Camera . . .
Saw wins.
And now . . .
It really does look like rain.

The starlight fades to black and—

INT. A POSH CORPORATE OFFICE—DAY

A big-money Vancouver office with a view of the Convention Centre and our tree-covered mountains. A powerful man in a suit rules the room, **JOHN CLEMENTS**.

INTERVIEWER (O.S.)
How do you feel about trees?

JOHN

I love trees.

When I was a boy, a tree saved my life.

I don't tell many people that, but I think for this . . .

When I was a boy I was coming home from school and a storm whipped up, out of nowhere. Thunder. Lightning.

It's funny, when I was a boy it was the thunder that scared me. Not the lightning . . . didn't like the noise.

I wanted to get home fast so I did just what I knew you weren't supposed to do. I cut across a field. And I ran. Because the thunder was chasing me. Because I was late for supper.

And there I was—running—in the middle of this field— right next to a cluster of apple trees. And I heard the thunder. Louder than I'd ever heard it before. And then I heard a noise, that scared me more than the thunder ever had.

A crash.

And a flash of the brightest light I've ever seen. Pure white light. And when my eyes could finally focus again, I saw it. An old apple tree . . . a few feet away from me, split, like an axe cut it straight through.

You could see the smoke. But it was more like steam. And it was sizzling, hissing. Bitter. Sharp. Like burnt apples. And I looked at the tree. And for just a moment I could've sworn—could've sworn it looked back and said—"That was for you. I took it for you."

I had this image in my head of those guards—the ones who stand around the president—

The ones who jump in front of him if bullets start flying. And I felt like . . .

Like the tree had done that for me.

And I stared at the stump and now I didn't feel the rain or see the lightning flashes or even hear the thunder.
I looked at that old tree, with the steam hissing from its core, and I said, "Thank you."

John's now framed by the industrial city view. His jacket is off.

INTERVIEWER
When did you start working in the woods?

JOHN
When I was still a boy.
Not much older.
I took a job on the green chain.
I helped sort the wood.
It was a good job. I was making more money than my father'd ever seen. Wasn't sure he approved—he never liked *the company*. But he didn't mind the money. I didn't like being inside, though. I missed the air. The trees. So I started working as a faller. Take a look—

He flexes.

JOHN (CONT'D)
Trust me—you work with an axe as long as I did—muscle never really goes away. Even when we switched to the machines, I still kept up with the axe. To keep in practice. Keep in shape. Even did one of those logging shows once. Really liked log-rolling. And man, could I take a tree down fast. Just wanted to let you know I didn't always work in a place like this. I worked in the woods. That's how I got here. That's why this matters. It's not just a job . . . It's my life.

It's always been my life.

There was a lot of—I guess you'd call it controversy—back then. I got tired of the fights.

Same song over and over.

Got sick of it. How do you keep fighting—for generations? For centuries.

How do you—why not just—move on?

Land claims, title claims, whatever you want to call it. The names change with the times and the lawyers and the PR people.

Do you know there's no such thing as the "Great Bear Rainforest?"

Some Greenpeace type just knew that if they gave the land a cute name it'd be easier to raise money.

Surprised they didn't call it "cuddles" or "fluffy." Guess that's next.

One thing out there the tree-huggers didn't have to invent.

The Spruce. The Golden Spruce. K'iid K'iyaas.

You know about this, right?

You're doing this, you got to know about it.

There's a tree. A sacred tree. K'iid K'iyaas. A Sitka spruce. Three hundred years old. A hundred and sixty-five feet tall. Sacred. The Haida say the tree holds the spirit of a boy who disobeyed his grandfather. And so he transformed into this tree. And the branches, the ones in sunlight. Gold. Like something out of a fairy tale.

John at the desk, framed by the mountains.

INTERVIEWER

You saw it. What was that like?

JOHN

I saw it once.

At a conference.

It's hard to . . . I'm sorry . . .

One of the wonders of the world.

Magic.

I never believed in spirits—never thought much about God—only really believed in what I could see. What I could touch. But that—I'll tell you—there was no way to look up at that—a tree that had been here before there was a Canada. Before there was an America. Before anyone talked about claims.

And not feel . . . connected . . .

Not feel . . .

Couldn't even take a picture of it. There was no way to get far enough back to fit it in the frame. I touched it though—touched the bark that had been there when my grandparents' grandparents' grandparents were babies. And I picked up a golden needle that had fallen on the grass. And as I held the golden needle between my thumb and my finger I could hear . . . Saw blades biting into spruce a few hundred metres away. It was the way the wind was blowing. The sound from the logging stand was carrying.

At the conference everybody talked about the claims, the case. I didn't say much. Just listened. I was there to provide . . . a presence. For the company. And when I talked later—outside the hall—I talked about the company. People wanted to know how *I* felt. What *I* felt. And I talked about the company. Our position.

This isn't just about money—or the environment. Everyone wants to bring it down to that. Everyone wants to make it about big business. And trees. But it's not just about that. It's about people—lots of people. Lots of lives. Families. Children. Communities. You can't just stop—change everything—people will suffer. They are suffering. Everyone wants it to be simple. It's not.

Some of the Greenpeacers just want to hand everything back—they think logging will just stop. Years of poverty. Hundreds of years of poverty. And they think if the claims go through, all the mills will just close. And we'll never see another clear-cut. Never occurs to them that some of us—

This company is trying to manage the forest—

The resources.

Yeah, like the politicians say—

mistakes were made—

But we learned.

We're doing this right.

I know these people.

They're my friends.

I believe that.

John at the desk, gets up walks over to a bonsai.

INTERVIEWER

Tell me about the concert.

JOHN

At that first concert, the "event"—the one a few years back, where all the big stars flew in from the States to tell us to leave the forest alone—

the company sent me.

"To put a face on our position." That's probably how you saw me, right? Why you wanted me to talk. I know.

I looked pretty shook up that day. Everybody thinks it was the concert, the speeches.

That wasn't it. It wasn't—that was the day—the Spruce—the Golden Spruce—K'iid K'iyaas—

A friend called—asked if I'd heard . . .

That was the day it happened. The day some crazy bastard cut it down. To protest logging, he said. Said if one tree was sacred, they were all sacred. Said he wanted to draw attention to what was happening—to the logging all around it. It was a statement.

A bullshit statement.

Like burning the *Mona Lisa* because you're mad about the cost of paint.

He got attention.

And if I could have—

I would have—

I'd still kill him.

He stole something. Something precious.

From me. From you. From my daughter.

And her daughter. And her daughter too.

It was supposed to live forever.

And I thought about that day.

I thought about the day the tree saved my life. When I was running across the field, trying to get home—

Back to the reserve—

Because the thunder was chasing me.

Because I was late for supper.

And the lightning hit.

And how that tree saved me and I said,

"How aa"—thank you.

And I thought about how—

Why—

The company wanted me to be their "face" for this.

Because if the Indian guy says it's okay to cut the trees . . .

But the thing is— I think the company—

I know they're trying—

And it's my home—

And if they go—

If they pull out—

People are gonna hurt.

I just wish . . .

He turns and looks out his window at the railway tracks below.

INT. DINER—NIGHT

Cleaning the diner we find **JENNI HOLM,** *a waitress.*

INTERVIEWER

How do you feel about trees?

JENNI

I love trees . . .

I love trees.

And everybody I ever loved loved trees.

My father—

My grandfather—

My husband, Ben.

Brett . . .

My brother, Dale.

My son, Jason.

They all loved trees. And yes, they cut them down.

Except for Brett. He was a grader.

And my son. He's at school. University.

And I'm not sure why or when or how that made them monsters.

How did men who live off the land—who love the land, died for the land—how—when did they become the bad guys? When did trees become like—like puppies? Or baby seals? They paid for my house. They paid for my car. For this—

She shows the ring.

JENNI (CONT'D)

And they're paying for my son to go to university—where Jason's learning to be a tree hugger and hate his dad.

My dad—

He started on the green chain.

Used to be you didn't have to finish school.

You could just work in the mill.

Used to be a lot more jobs.

Used to be a lot more trees.

But there are still enough.

Even with the fire.

These trees are our life.

You live here, you have to love trees.

And I'm tired of people coming from outside—

That actress was here—

You know her. Leila Cole. She was in that vampire series. She sings too. Never heard her on the radio—but you see her videos all the time.

And she was in the news because of that time she got arrested. And the thing with the photographer. And all the fights with her boyfriend. And she's here asking for a burger with no bun. And salad. No fries. And dressing on the side. And diet cola with a wedge of lime.

"Not a slice. A wedge."

"And just two cubes of ice."

And she's got all these other people with her.

And at her table—

(indicates table)

Right over there—this girl—taking notes. Doing an interview, I guess. And this big black guy.

Her guard maybe?

And this girl asked Leila where we should be cutting trees.

She didn't know.

But she said we should all recycle more.

The old lady was here too. Before she got arrested.

And the kid that's up in the tree. He's a vegetarian.

The kind that doesn't even eat eggs.

Just watch—it'll get windy one night. He'll fall.

And if he gets lucky—if he just breaks his neck—he'll sue the company—sue the town. Say it's our fault.

My husband . . . He tried to talk him down once. Told him about that storm. When there's gonna be a real storm you can feel it. So Ben warned him.

But you know what the kid did? Dumped shit on him. No lie. The kid does his thing in a bucket. And when Ben was talking to him, trying to warn him, the kid dumped the bucket.

Just missed him. Ben was so mad—he revved up the chainsaw—to give the kid a scare. And then—then he just gave up. But he's not going out there after any storm. He's seen enough bodies. Been to enough funerals.

When my dad went to funerals—for friends—before

he got old—it was because they died in the woods . . .
Accidents. A tree'd fall the wrong way. Something would
go wrong at the mill. Or they'd be out fighting a fire and
the wind would change . . .

Like Dale . . .

When our friends die—

it's not the forest that kills them.

Not the job though.

Layoffs.

Cutbacks.

Rationalization.

I loved that one.

They weren't fired.

They were "rationalized."

Some of them lost their homes.

Their trucks.

Three of them—

Andy just drank himself to death.

Allan. You heard about Allan.

Probably why you're here.

He made all the news.

He's the one who shot himself.

Went home. Shot himself.

But shot his kids first. And his wife.

Three kids. Left a note—

His youngest . . . She was five.

She was at daycare.

(beat)

There's a fund for her now.

She shows us a jar filled with a mix of bills and coins. There's a picture of a young girl taped to the side.

JENNI (CONT'D)

And Brett . . .

He was fighting fires when I first met him.

Fighting fires at 16.

How sexy is that?

We were together back then. Two years before—

Anyway, Brett got rationalized.

And there was nothing else. All he knew how to do was work at the mill.

Government said they'd retrain him. He could learn to type. Use computers. So after he was rationalized . . .

He signed up to fight fires again. And when he wasn't fighting fires—he was working at the Dairy Queen.

So when the fire started . . . I knew it was bad—but I started going to the Dairy Queen again. I know it's not great money. Fighting fires. Not mill money. But it's money. As long as there were fires.

So last summer—you know—Brett got a lot of work.

He looked good—

He'd come in here some nights—after he'd been out there—still covered with sweat. Said maybe he'd train to be a jumper. Back when we were kids he wanted to be a jumper. I wanted him to be a jumper.

He looked good.

But I guess it was tough to watch it all burn.

Because that's mostly what you do—

Watch it burn.

You can't stop it.

Just try not to let it spread.

And then the fires stopped.

And he stopped coming in here.
And then he just walked into the woods one night—
Not long after you talked to him.
Never came back.
There's no way it was his cigarette.
I don't know what people told you—but there's no way
he woulda done that. Not after Dale.
Police put together a search party.
Didn't find him.
So maybe he's still out there.
Maybe he just doesn't want to be found.
His wife had a funeral for him. A memorial.
Then she left.
Lots of people left.
Ben only got partway rationalized.
But at least he's still got shifts.
And that's when I started this.
I hadn't worked since I was 17.
And now here I am listening to this actress talk about
how much she loves trees.
I should've spit in her salad.
I should've told her to go back to LA. Told her—
But I just got her her burger with no bun.
And her salad, no fries. And her diet cola.
And we don't even have lime.
She gave me a big tip anyway.
She handed me a hundred dollars.
A hundred dollars. US.
For a burger with no bun.
And the black guy—the guard—just had a decaf. And a
slice of pie.
So the whole bill—eleven thirty-five—
And she told me to keep the change.

Biggest tip I've ever seen.

(beat)

I put it in the jar . . .
Know what I shoulda told her?
I thought about it after she left.
I shoulda told her what I told you.
I married a logger.
My dad was a logger.
We live off this land.
And we love trees.

We hold on Jenni until she looks away and we PULL BACK as—
EXT. FOREST—DAY

It's the end of the day and a full logging truck pulls away and down the road where Ben was cutting trees and—

FADE OUT

ACKNOWLEDGEMENTS

When I was in grade nine, the Eurasian milfoil weed had appeared in Okanagan Lake, and BC's provincial government decided to combat it with the chemical compound 2,4-D, a controversial herbicide. There were all sorts of stories in the newspapers indicating that the pesticides were going to be even more damaging than the weeds they were supposed to stop. My science teacher gave us an assignment to write about anything science related that interested us, so I wanted to write about 2,4-D. I wanted to find out more than what was in the newspaper but couldn't figure out how until my mom suggested, "Why don't you just call someone from SPEC?" (which then stood for the Scientific Pollution and Environmental Control Society, and later became the Society Promoting Environmental Conservation).

So I did exactly what any grade nine student would do. I looked up the number, called and spoke to the chairperson in charge of fighting 2,4-D. I can't remember how long Merriam Doucet spoke to me, but she informed me that "some of the chemicals contained in 2,4-D were used as poisons in Vietnam" and that these poisons could seep into the drinking water.

I'm not sure whether I was actually more concerned with the drinking water or the Ogopogo (an Ogopogo expert had publicly warned that 2,4-D might kill the lake monster). But I do remember that when I presented my paper, my teacher was shocked that I'd actually interviewed a real live expert. And I got an A+.

After that, whenever I wanted to find out something about what was wrong with the world, I'd try to find someone who had the answers and ask them. Thanks, Mom!

These particular interviews exist because growing up in BC I really wanted to know what was wrong with our forests and what should be going on in our forests and in forests around the world.

Many of these conversations were originally part of my podcast series *Trees and Us*, which were and still are available from *The Tyee* (www.thetyee.ca). *Trees and Us* would not have been possible without the support of *The Tyee* and founding editor David Beers—and Kat Dodds, who urged me to create the podcast and said, "Just watch, one day it'll be a book too."

These interviews may never have ended up in print if Sam VanSchie hadn't agreed to transcribe and fact-check them. I'd also like to thank the amazing women who checked out the early drafts for me—Rayne, Deanne Beattie and especially my editor, Heather Sangster, who had the challenge of helping me selectively log this forest of interviews.

A million thanks to everyone who agreed to be interviewed for the book and the podcast series. It broke my heart to cut so many of your words, but I hope these interviews will steer people to your books, organizations and other projects. I'd especially like to thank John Wiggers for his passionate support of every aspect of my *Green Chain* adventures, Tzeporah Berman for steering me to some of the interviews, and Harry Bardal for the chilling and stunning cover image (and our movie poster) that was created from composite photographs he took of storm-damaged trees in Stanley Park.

And *mahalo nui loa* to everyone who made *The Green Chain* movie happen. It's a long, long list—although you can meet most of the Green Team by visiting www.thegreenchain.com. But I have to mention the original Green Team—Tony Wosk, Donna Wong-Juliani and Darron Leiren-Young, who now joke that I only made the movie so that I could do the podcast and the book.

And thanks to Heritage House and the passion of Vivian Sinclair for encouraging me to create this book you're holding right now. For more on all the links in *The Green Chain*—and to listen to many of these interviews in their entirety—visit www.thegreenchain.com or connect through my site at www.leiren-young.com.

Now go outside and hug a tree . . .

ABOUT THE AUTHOR

Mark Leiren-Young is the bestselling author of *Never Shoot a Stampede Queen*, winner of the 2009 Stephen Leacock Medal for Humour.

Mark is also one of Canada's "greenest" writers. He wrote, directed and produced the award-winning feature film *The Green Chain;* wrote and coproduced the gold medal-winning short comedy *The Green Film;* and wrote, produced and starred in the EarthVision award-winning TV special *Greenpieces*—the world's first eco-comedy. His satirical comedy duo, Local Anxiety, has been featured on CBC and NPR and has released two albums, *Forgive Us We're Canadian* and *Greenpieces*.

His stage plays—none of which currently include the word "green" in their titles—have been produced throughout Canada and the US and seen in Europe and Australia. Mark has written for such publications as *Time, Maclean's* and *The Utne Reader*, and he's a regular contributor to *The Georgia Straight* and thetyee.ca.

Mark was born in Vancouver and earned his BFA in theatre and creative writing from the University of Victoria. He currently splits his time between Vancouver, BC, and Haiku, Maui.

For more on Mark, visit his website at www.leiren-young.com.

Also by Mark Leiren-Young

Never Shoot a Stampede Queen
A Rookie Reporter in the Cariboo

True-life tall tales from a rookie reporter's adventures
in Canada's still-very-wild west.

Winner of the 2009
Stephen Leacock
Medal for Humour

"Funny, moving and profound. You will laugh out loud."
—Will Ferguson

"*Never Shoot a Stampede Queen* isn't just sound advice, it's also the
most fun I've had this year." —Spider Robinson

"[E]ach tale grabs my attention . . . The Cariboo has never looked
so . . . dangerous!" —John Robert Colombo

"Gun fights, bar fights, plane crashes and, of course, mad bombers, all
as seen through the fresh eyes of a newbie reporter." —*Vancouver Sun*

ISBN 978-1-894974-52-3
www.heritagehouse.ca